RETHINKING THE REGION

John Allen, Doreen Massey and Allan Cochrane
with Julie Charlesworth, Gill Court, Nick Henry and Phil Sarre

London and New York

First published 1998
by Routledge
11 New Fetter Lane, London EC4P 4EE

Simultaneously published in the USA and Canada
by Routledge
29 West 35th Street, New York, NY 10001

Typeset in Goudy by Keystroke, Jacaranda Lodge, Wolverhampton
Printed and bound in Great Britain by Biddles Ltd, Guildford and King's Lynn

British Library Cataloguing in Publication Data
A catalogue record for this book is available from the British Library

Library of Congress Cataloging in Publication Data

ISBN 0–415–16821–X (hbk)
ISBN 0–415–16822–8 (pbk)

CONTENTS

List of maps and montages vi
Preface vii

Introduction: A space of a neo-liberal heartland 1

PART I
Discontinuous regions **7**

1 When was the south east? 9

2 Where is the south east? 32

PART II
Regions and identities **63**

3 Identity of places 65

4 Spaces of identity 90

PART III
Space–times of neo-liberalism **115**

5 Self-defeating growth? 117

6 Space, place and time 137

Bibliography 144
Index 153

MAPS AND MONTAGES

Maps

2.1 The south east Standard Region 33
2.2 Distribution of earnings: females, 1989 36
2.3 Distribution of earnings: males, 1989 37
2.4 Distribution of high incomes (over £30,000 per annum), 1989/90 38
2.5 Distribution of high incomes (over £40,000 per annum), 1989/90 39
2.6 Distribution of high incomes (over £50,000 per annum), 1989/90 40
2.7 Regional change in earnings, 1979–86 41
2.8 Regional change in house prices, 1979–86 42
2.9 Finance: employment, consumption and income reach 43
2.10 High-tech employment 44
2.11 Regional change in national government spending: roads and transport, 1987–91 45
2.12 Regional change in national government spending: regional spending per capita (including defence), 1987–91 46
2.13 Regional change in national government spending: housing, 1987–91 47
2.14 Regional change in national government spending: trade and industry, 1987–91 48
2.15 International financial links 49
2.16 Local project areas 59
5.1 Traditional heartlands and growth areas of the European Community 121

Montages

1.1 The establishment culture of 'the white ROSE' 30
3.1 Hot spots and 'holes' in the doily 69
4.1 Entrepreneurial masculinities 96

PREFACE

The spectacular defeat of the Conservatives in the May 1997 general election is testament to one of the central claims of this book: the unsustainable nature of growth based on market values, entrenched inequality and strident individualism. What the election showed was the inherent fragility of the neo-liberal project, as the electorate, even in its heartlands, was no longer prepared to shoulder social division and insecurity as the price to be paid for economic growth – the fruits of which were always never quite within reach. The electoral map of the south east changed overnight, with larger than national swings reducing the Conservatives largely to the rump of the home counties, their traditional core before the neo-liberal shift. As an agenda of political and cultural change, the legitimacy of neo-liberal growth has not only been called into question by the electorate of the south east, it has been dramatically cut away geographically.

Ironically, it is the areas that they still hold in the south east which do now meet their much vaunted image of the nation – the rural idyll and its picturesque English 'villages'. With the reorganization of constituencies, the Conservatives lost many of the middle-sized towns in the south east – Bedford, Harlow, Hemel Hempstead, Luton, Milton Keynes, Stevenage, Watford to name but a few. At the same time, they retained the rural and suburban seats in Buckinghamshire, the likes of which perhaps says more about what is quintessentially Tory than many other locations in the region. Although not absent from the town and cities in terms of individual votes, the Conservatives are perhaps now giving real electoral credence to the acronym – ROSE – as the *rest* of the south east.

More importantly, such a shift opens up the possibility for a new south east, in both representational terms and in terms of a new regional politics. In this book, we argue for the possibility of an alternative political project: one based upon the livelihood and fortunes of those who live in the south east, as much as those living beyond it. The recognition of a more balanced, even form of growth which recognizes regional differences within the wider, national and international context in which they have been formed represents just such a potential starting-point.

Of course, new political settlements are not put in place by merely questioning the validity of regional or institutional boundaries, nor by willing a less divisive social order based on solidarity, morality and trust. The political and cultural legacy of the south east is one very different from that associated with Labour, even in its 'new' political colours and trappings. The issue is not one of attempting to turn the clock back to the social democratic conditions of thirty-odd years ago, simply to re-create the growth dynamics of the 1960s. Rather, the issue, as stressed in various ways throughout this book, is to re-imagine the region in an open, relational political context. In short, we need to rethink the region, politically and geographically.

Rethinking the Region is the outcome of a collaborative effort which started in October 1989 as a discussion group among geographers at the Open University, together with Nigel Thrift and Andrew Leyshon from the University of Bristol. The practice of working in course teams at the Open University to produce teaching materials provided us with a model way of working which we were keen to extend to the research process. Having produced a six-volume course on Restructuring Britain in the latter part of the 1980s, it was felt that the practice of teamwork, and indeed many of the ideas developed on the nature of social transformation, could usefully inform our research thinking and practices. The initial sparks which kindled our imaginations and brought us together were a political concern with the nature of the growth laid down in the 1980s, especially its deep inequality, and a strong commitment to thinking through the geographical significance of growth in what, to all intents and purposes, had become known as the heartland of neo-liberalism – the south east region. Sitting in Milton Keynes, we were acutely aware that growth could take many different forms. And indeed results from the ESRC-sponsored 'Localities' projects, with which we interacted productively, provided constant stimulation and provocation in this and many other respects.

The nature of neo-liberal growth, with its markedly uneven and divisive effects prompted us to conceptualize the different axes of growth involved and from there to identify their points of geographical concentration, as well as those locations where growth in the region was largely absent. Indeed, such an endeavour prompted us to ask exactly when and where was the south east? The outcome of this collective endeavour was the construction of a broad research framework which was sufficiently robust as to enable researchers within the groups to study in greater detail different sides to neo-liberal growth. Funding was sought, and obtained, from the ESRC for a two-year programme of research comprising four linked projects, each designed to reveal aspects of the cross-cutting nature and varied impact of growth across the south east. The rest, as they say, is geography. Or, rather, the rest, the collaborative output of the programme, is presented in this text, and in a wide range of published articles and papers to both academic and non-academic audiences.

We should stress that this book is not a research monograph in the traditional sense of the term. Our major aim was to *conceptualize* the geography of growth, its character and potential fragility, and in so doing to reveal how regions are made and remade over space and time. For those seeking a comprehensive, statistical picture of the south east as a growth region, this book will fall short of their expectations. The book does, none the less, contain a wide range of empirical material gathered from the linked projects, and those wishing to pursue the detail of the findings presented here will find the Occasional Paper Series associated with the programme of research (listed at the end of the bibliography) of considerable interest and value.

The biggest problem we wrestled with from the very beginning was 'What was the south east?' The issue, which raised itself in meeting after meeting, was that any attempt to map the social relations of growth, to capture their social reach as well as their points of overlap, ran the risk of misinterpretation. On the one hand, there is an obvious need to ensure that the south east is recognizable as an object of study and that we are all in fact talking about the same thing, initially at least. The boundary lines that govern the physical and political relationships of the region, for example, may provide such a common understanding. On the other hand, to draw such precise boundaries is to invite misunderstanding. Once drawn, such lines of containment convey the impression that all the social relations relevant to an understanding of growth in the region fall neatly within the boundaries. The result, effectively, is to empty the region of meaning and to fix its changing geography. The maps of the region in this book, therefore, give only an indicative impression of the complexity we have in our heads. None the less, we hope that we have conveyed something of the overlapping, intersecting relationships which are involved in the making of a region. It is for similar reasons that we have referred to regions throughout in lower case rather than their formal upper case notation.

Finally, it remains to acknowledge those who have contributed to this programme of research over time and to detail responsibilities for the final version presented here. It is no exaggeration to say that the contributions of those who have now left the Open University, Chris Hamnett and Linda McDowell, as well as those from the University of Bristol, Nigel Thrift and Andrew Leyshon, were critical to the original conception of the project and to its development into a research programme. Chris Hamnett's understanding of the workings of residential and commercial property markets was an invaluable asset, as were Nigel Thrift and Andrew Leyshon's insights into the operation and impact of financial markets across the south east. Similarly, Linda McDowell's grasp of gender, sexuality and identity strongly influenced the research framework, and the results of her own work as part of the programme are published in *Capital Culture: Money, Sex and Power at Work* (1997).

Naturally, all of the above are exempt from responsibility for the final version of the text. All seven authors were fully involved in discussions of draft chapters, although final responsibility rests, in the case of Part I, with John Allen and Doreen Massey; for Part II, with Julie Charlesworth, Gill Court, Nick Henry, Phil Sarre, John Allen and Doreen Massey; and for Part III, with Allan Cochrane and Doreen Massey. The maps were produced by John Hunt and Jenny Seavers, and Margaret Charters and Doreen Warwick provided secretarial support. Edward Hall painstakingly put together the photomontages. To them, we owe our special thanks.

INTRODUCTION

A space of a neo-liberal heartland

One of geography's central objects of study is the region – the place, the specific area. In 'the old days' textbooks on regional geography worked their stately progress through a sequence of regions which seemed to have existed for all time (the eastern seaboard, the Appalachians, the midwest . . .). There was subsequently a period of dominance by quantitative studies and model-builders, which demanded above all consistency and the clarity of boundaries; statistical regions became the requirement (in the UK the 'Standard Regions': the south east, west midlands, east anglia . . .). More recently we became enmeshed in the continuing debate about localities and locality studies – how to define them and how, even whether, to do them. The question of 'place', at various scales and in various guises, is a widespread topic of concern. And so too are questions of boundaries, borders and spatiality more generally. All this has raised issues of theoretical approach, and even of the nature of theory itself; of the conceptualization of places and of their practical definition; and of what should be studied 'within' them. Today these questions are at the centre of the agenda of economic and social geography, as well as integral to developments in social and cultural studies. While debate goes on, studies continue to be produced from Los Angeles to Lancaster and from Sydney to Sheffield.

This book is an intervention in that debate. It accepts the validity of place-specific studies: as exemplars of wider phenomena, symptomatic of broader changes; as laboratories for the exploration of particular issues, both theoretical and empirical; and for themselves, to aid attempts by people living and working within an area to understand what is going on around them and maybe to change it for the better. But this book is also an argument for a different way of approaching this kind of regional geography.

It argues, most particularly, that the manner of conceptualizing a region is intimately bound up with the wider debate about the conceptualization of space and place. If our approach to defining a region can be summarized, it is probably best captured by two principles. First, it embodies a strongly relational approach to thinking about space and place. That is, it understands both space and place as constituted out of spatialized social relations – and narratives about them – which not only lay down ever-new regional geographies, but also work to reshape social and cultural identities and how they are represented. Second, it acknowledges that such studies are always done for a purpose, with a specific aim in view. Whether theoretical, political, cultural or whatever, there is always a specific focus. One cannot study everything, and there are always multiple ways of seeing a place: there is no complete 'portrait of a region'. Moreover, we want to argue, 'regions' only exist in relation to particular criteria. They are not 'out there' waiting to be discovered; they are our (and others') constructions.

Our immediate empirical purpose in the research presented here was to explore the dramatic growth which took place in the 1980s, the contradictory effects which it produced and which were all too often ignored, and the legacy of changes which confront us today. The 1980s were the heyday of 'liberal' free-market capitalism. While this took very different forms in different countries it was clearly an internationally recognizable phenomenon. There was Reaganism; there was Thatcherism; from Spain to New Zealand even nominally 'labour' or 'socialist' parties adopted the approach with gusto; the IMF and the World Bank lent their energies to exporting it to the third world. Whether or not it was a 'success' depends on your criteria and where you stand. Certainly, it created upheavals and inequalities. Some of these have been apparent for a while in poorer areas (the riots which so often greeted IMF programmes of structural adjustment being one index of this, as was the devastation wrought in the early 1980s upon manufacturing industry in the north and west of England, for instance) though even this is only slowly penetrating official consciousness. What this study does, however, is look into the contours of that growth within one of the reputedly most *successful* regions: the south east of England. This place was both arena and icon of the neo-liberal project. Here, it is said, neo-liberalism indubitably succeeded.

But did it? Or, more precisely, and of more relevance to a critique of the neo-liberal project as a whole, what were the terms of its 'success'? Our aim, in other words, is to enquire into the nature of the neo-liberal form of growth in one of the international heartlands of its triumph and through this to illuminate a particular construction of a dominant region.

The success of the south east of England, indeed, became one of the most powerful elements in the imaginary geography of the United Kingdom in that decade. This was an imaginary geography held in different forms by many people across social groups and across political perspectives. What all agreed upon was that the country was divided, and increasingly so, between the south east and the rest: the south east was the growth region; the rest of the country was in decline. The characterization of this geographical divide, and the reasons adduced for it, of course varied widely. For those on the right of the political spectrum, the south east was proof that the new market philosophy could succeed: here was entrepreneurialism, here were sunrise industries, here there were few trade unions . . . In 'the north', by contrast, people were still mired in corporatism, weighed down by labourism, trade unions, sunset industries . . . The left, of course, saw it differently: saw the south east as a land of yuppies, greed, fictitious capital and spiralling house prices, where the triumph of individualism simply meant selfishness. The north, for them, was where old values and verities hung on. However one saw it, the divide was powerful. It was reflected politically, and it was mobilized by Thatcherism in its proselytizing mission. Other regions were urged to become more like the south east, to become more entrepreneurial, to abandon the supposed constraints of collectivist philosophy. So growth in the south east was important in that decade, and it retains a significant symbolic resonance. What we are concerned to explore are some of the complexities of this picture, and the conflicts and contradictions within it. To look inside a region which triumphed, or was widely understood to have triumphed, on 'free-market' growth.

More than that, we are concerned to show that we now face the consequences of a form of growth built upon individualism and market values. Put simply, regions are not given to us; they come with a particular set of economic and social legacies and a particular geography. In a very real sense, they set the context for how we can re-imagine the south east and its relations to the rest of the UK and its regions.

This, then, is a story about the nature of the south east of England as a growth region in the 1980s and about the legacy of changes which confront us today. But woven into it and emerging from it is a host of more general arguments. This was, as we have already indicated, a decade of growth of a very particular sort. In the UK, the growth of the 1980s was built on finance and professional services, on credit, consumption and house prices, on deregulation and privatization. On the other hand, it led to or was associated with sharp increases in social inequality in a variety of dimensions, both in the UK as a whole and within the south east itself; it led to increasing geographical inequality, in particular in the shape of the north–south divide in the UK, and more local inequalities within the south east itself; it led to the reworking – often painful, sometimes contradictory – of social relations and identities, in particular around class, gender and ethnicity; and it has proved to be fragile as a result of its very composition. 'Growth' can take particular forms, and each form will have specific strengths and weaknesses, different distributional effects and different geographical, social and cultural implications.

To look inside the south east at a moment when the region was pulling away from the rest of the country on the basis of 'free-market' growth is not, as we have indicated, to construct an exhaustive profile of the region. Rather, it is to focus on what defined the place as successful, and successful in a very particular way. In our view, the peculiarity of the region at this time is best understood as the outcome of a particular bundle of dynamics – dynamics, or mechanisms, of growth – which, in combination, reproduced the dominance of the south east in the 1980s – this time in neo-liberal colours.

The nature of these dynamics, the axes of growth through which the south east has maintained its hegemonic position in the UK, will be spelt out in subsequent chapters. For the moment it is important to note, first, that taken individually, the identification of these different components of growth – the financial services boom which gave rise to employment gains across the south, the explosion in credit and borrowing which fuelled household consumption in the region, the spiralling of house prices and the boom in commercial property markets which erupted in places like London's docklands, the burst of high-technology growth in the south, and the massive public expenditure on infrastructure in the region, as well as the stimulus of the governments own policies on privatization, deregulation and so forth – does not represent and is not intended to represent a series of novel claims. Indeed, each of these dynamics has been outlined and discussed at length by various commentators. Rather, our concern here is with the ways in which, at a particular moment, they came together to produce a 'neo-liberal region' with its own distinctive geography. These dynamics touched particular places and particular groups, parts of 'the region' were included and others excluded, frequently the dynamics reached beyond the formal boundaries of what we had thought of as the region – in other words, these interactions *constructed* the south east of the 1980s. Within a particular time-frame they defined the region. It is this that represents the basis of our approach to the region and it is this relational way of thinking about the region which, we would argue, represents a more adequate understanding of how the social relations of places are made and remade.

A second feature of our approach is that it is not our intention to measure the distribution of growth or to provide an economic assessment of regional growth dynamics. This study is of a rather different kind. Our main concern is with the axes of growth, with the nature of the growth mechanisms themselves – the propulsive aspects of growth rather than simply the distributional benefits or spin-offs from, say, the expansion of the financial services sector. So in terms of the consumption-led boom of the 1980s, for example, our concern was not simply with the vast range of purchases made, but with the specific items of consumption which are themselves distinctly propulsive, such as the purchase of property or other high-value items. Taken together, the spatiality of the dynamics, the mechanisms of growth and their uneven geography, represent a particular way of conceptualizing the south east as the 'growth region' of the 1980s and, indeed, of conceptualizing places more widely.

This book is also, as stressed above, about the 'effects' integral to that form of growth. So, having in the opening chapters wrestled with the concept of region and the nature of growth, we go on to look at how this particular neo-liberal growth, and the forms of social and geographical inequality inscribed within it, generated new lines of difference and division – both social and geographical – from those previously dominant in the region. Moreover, as the historical homeland of 'one nation' Toryism, the inherited south east of the home counties and the City of London clashed with the strident nature with which free-market values and competitive enterprise were promoted within the neo-liberal project. In the creation of new inequalities, and in the negotiations between established and emergent social groups, gender, ethnicity and class (and the spatialities of each of these) were all at stake and all emerged in some way or another reworked.

At different times, the space called the south east has been shaped by a variety of forms of growth, each with the potential to remake social relations across the region. The aim of this book is to capture that process in the 1980s when what was happening in the south east was emblematic of the country as a whole and, in fact, emblematic of an approach to economic growth that had attained global hegemony. One of the wider messages of this book is therefore to underline the damaging and enduring effects – even in its triumphant heartlands – which, we would argue inevitably, it produced, and the ways in which it indeed continues to shape the production of new social and geographical identities. But that message is itself set within a wider conceptual argument: that we must rethink the way that we approach regional geography. In arguing for a relational approach to space and place, the interpretation set down here of a changing south east develops that approach and highlights its broader applicability. Conceived as a series of open, discontinuous spaces constituted by social relationships which stretch across them in a variety of ways, this book represents a starting-point for an alternative way of thinking about the region.

Part I

DISCONTINUOUS REGIONS

1

WHEN WAS THE SOUTH EAST?

Spaces/places are constructed both materially and discursively, and each modality of this construction affects the other. Moreover, every place or region 'arrives' at the present moment trailing long histories: histories of economics and politics, of gender, class and ethnicity; and histories, too, of the many different stories which have been told about all of these. The complex ways in which a region is constructed and read at any time is a result of these histories and of what is made of them. In the 1980s, the south eastern parts of the United Kingdom clearly had the status of being the dominant region of growth in the country. It was a status which built upon an historical inheritance both material and discursive. Yet, so we argue, economic growth can come in many varieties and each will have different effects. The growth through which the south east evolved in that decade was of a highly particular kind – 'Thatcherite', 'neo-liberal' – and, while drawing on the inheritance of the past, it was to change yet again the nature of the place and its social construction. What was created in the 1980s – as any region anywhere is constantly re-created – was a new constellation of forces.

For many commentators, the type of growth that brought the south east region to the fore in the 1980s lay at the core of the neo-liberal project. In itself, the region was emblematic of a broader political project which had a particular 'take' on growth. This was not just a project which exhorted 'more growth': more jobs, more investment, more homes, more money in peoples' pockets and other such measures; it was a project about growth of a particular kind. It was a project which comprised a cocktail of elements, not all intended, which, when the chemistry worked, produced a 'south east' along new lines of difference and inequality, both socially and geographically. As a 'growth region', the south east was the site for a remaking of social relations along neo-liberal lines, according to a particular blueprint of 'success'. Success in this context meant individual success. It meant depending upon self-reliance, personal ambition, an ethic of hard work and the ability to take advantage of what opportunities came your way in a competitive environment, with little or no concern for the inequities involved. And in the 1980s nowhere in the UK encapsulated this particular image of growth more than the south east of England.

In this chapter, we look at the manner in which the presentation of the south east as *the* UK growth region of the 1980s became one of the ways in which it was possible, not only to 'read' what was going on in this part of the UK, but also to see a political project at work. In the 1980s, the regional identity of the south east, indeed its very dominance as a region, took its meaning from the fact of growth – of a neo-liberal kind. Such a representation, however, did not simply fall into place; it had to be produced.

Representing dominance through growth

As a representation of the south east, the notion of it as a 'growth region' did not take hold without there being certain legacies of meaning already in place. The identities of regions are constructed through their relationships to 'other' regions and naturally they come with a history in which they have already been 'placed', so to speak. By 'placed', here, we mean that regions draw their meaning at any one point in time through their differences from other regions. They are already inscribed with meaning. They are part of a *system of representation* which, among other positionings, refers to 'core regions', 'peripheral regions', 'manufacturing regions', 'wrecked regions', 'poor regions', 'high-tech regions' and the like. The regional identities are relational, marking out the differences and contrasts between regions, and, whilst they are open to reinterpretation, they carry a legacy of meaning. It would jar for example, to refer to central Scotland or east anglia as core or lead regions, because the resonances of the past, the regional identities attached to them over the past century or so, render such labels implausible. It is not that the identities of the two regions are fixed; they can and do shift, but only as part of an historical play of differences between regions. Not all regions can be core at the same time, although they may take a variety of other designations.

By the same token, it was possible, in the depths of economic recession in the late 1980s and early 1990s when the south east bore the brunt of the collapse, still to refer to the region as a core region. It was widely assumed that, of all the regions in the UK, the south east was best positioned to lead the national economic recovery when it came. This attribution was not simply because the south east is by far the largest region in the UK, in terms of its wealth, the size of its work-force and other distributional characteristics: it was also because for many people the south east continues to be the dominant region of the UK. As part of an imaginary regional geography of the UK, the south east holds pole position in a *discourse of dominance*.

It is important to stress here that there are powerful images in play in such a discourse, and they did not take shape overnight. Regardless of the economic fortunes of the south east – whether it is or is not growing – the region has long occupied a dominant position, economically, politically and culturally within the UK. This was the view expressed by Breheny and Congdon (1989) and, although we shall later question their formulation of this dominance, it is not particularly difficult to understand why they and others may easily draw such a conclusion.

A 'two nations' saga – the divide between an old industrial 'north' and a prosperous commercial 'south' – has been part of popular discourse in the UK for the best part of a century or more. Within such a discourse, however, the 'south', and the south east in particular, is already *over*endowed with meaning, some of which is contradictory in its signification.

Smith, in his account of the growing 'north–south divide' at the end of the 1980s, listed a number of grounds for the dominance of the south east and the crucial role of London:

> A prime cause of the southern bias of the British Society, is the concentration of decision-making power in London and the south east. London is the seat of government and the home for a huge civil service machine. The financial system is centred upon the City of London, to the virtual exclusion of all the regional financial centres. Most company head offices are in London and the south east, spawning a vast array of support activities in administration, advertising and marketing, research and development, and financial and business services such as banking, accountancy and management consultancy. The national media, other broadcast and print, are exclusively London-based. National opinion-formers, if that is not too grand a term, see things from a south eastern point of view, and attempt to influence national decision-makers, who are subject to the same regional bias.
>
> (Smith, 1989: 213)

Reflecting on this account at the end of 1991, in the wake of a full-blown recession in the south east, Smith (1993) saw no reason to alter his view. Having gone from 'boom to bust' – the title of his later book – the south east region was still considered by him to occupy a dominant position in the UK economy. There is, he argued, a long-term trend in favour of the south on the grounds of the above fundamental characteristics, plus the 'pull of Europe' and the sheer concentration of wealth in the region, which together can be seen to promote economic growth.

Regardless of what we think of this line of argument, it is possible to see how the above series of characteristics is capable of reproducing the dominance of the south east region within a system of representation. The south east remains the 'core region' because it continues to host an international centre of finance, because it continues to be the region of government, of media, of corporate services and so on. Conversely, other regions are not characterised in precisely this way, even though – for example – Scotland can lay claim to certain decision-making functions, most notably in the field of legal and educational issues. Thus, the strength of the imagery is as much about what other regions are not, as it is about what the south east is. More importantly, there are a number of different ways in which the distinctiveness, the very dominance of the south east, may be expressed. Not all accounts of regional dominance, for example, need refer to the economic significance of company HQs or the concentration of international finance in the region. The dominance of the region may also be expressed in class terms and culturally. Thus, in the case put forward by Weiner (1981) and others, the anti-industrial governing elites of the Victorian era in the UK laid the basis for an ideology of 'gentlemanly' dominance which was centred on London and the home counties. Or again, the stress may be placed upon the location of political power in the region and the concentration there of state administrative elites, or dominance could be expressed in terms of a more general political hegemony, as was the case in the 1980s. Clearly, the lines of justification may overlap, but what matters is that all accounts, despite broad variation in the characteristics or relationships identified, support a common line of thought: namely, that the south east is *the* dominant region of the UK.

To suggest as much, however, is not to produce a fixed image of the south east region. As part of a discourse of the regions, the dominance of the south east has been constructed over time through a variety of statements and elements drawn from other discourses (discourses of social class, of cultural capital, of political elites and of geographical location in relation to Europe). At any one moment, therefore, if a particular interpretation of the south east is to be sustained, a regional discourse is likely to incorporate and adapt new elements so as to appear plausible in the face of changing events . It is not that the south east lacks the material basis for its dominance, but rather that the form and representation of that dominance have shifted over time as part of a contested process. The 'pull of Europe', for example, is one such element, drawn upon by Smith to reaffirm the south east's pole position at a time when its economy was 'bust'. Similarly, in the mid-1980s, when the south east was in a 'boom' phase, the success of the region was frequently described in terms of its moral culture and the entrepreneurial dynamism of certain social groups. In particular, an enterprising professional and managerial stratum in a range of private services, from banking and finance to leisure and cultural services (which comprised the new middle class), was identified with the success of neo-liberal forms of growth. The iconic figures of success in the south east of the 1980s – shareholders and homeowners, credit-card holders and mobile-phone users, drivers of BMWs and dealers in stocks and shares – were largely different from those of previous periods. Both during the 1980s and before, the stress upon a particular social grouping, or the emphasis placed upon proximity or economic and political connections to Europe, reinforced the idea of the south east as the dominant region. The different elements were drawn into the discourse of dominance and woven into its network of meanings to support a particular interpretation of the region. In the 1980s, the south east reproduced itself as the dominant region in the UK in part through attempts by the neo-liberal right to project it as *the* growth region. The south east region has manifestly experienced specific eras of growth before, in the 1950s for instance, but in the 1980s its dominance was represented through (this particular form of) growth, through new social groups and institutions, and along new lines of social difference and inequality.

Let us be clear what is being argued here. The south east *is* in many ways the dominant region in the UK. It is so materially (in class, economic or political terms, for instance) and it is so discursively. The two modes (material and discursive) are largely mutually reinforcing. However, the way this works – both materially and discursively – varies over time. What we shall see in the case of the south east in the 1980s is, first, how the Thatcherite project seized upon this history and both wove a new element into this historical discourse of dominance (the region as home of dynamic entrepreneurship for example) and materially reinforced its dominance over other regions. In other words, the neo-liberal project certainly had something to build on, but it also re-moulded what was found to hand. However, we shall also find – second – that this process had its own contradictions. In the remainder of this chapter, we shall look at the overlapping narratives of dominance and growth which lie behind the notion of the south east as a 'growth region' and point towards the ways in which a particular form of neo-liberal growth remade social relationships across the region in the 1980s and early 1990s.

Narratives of growth

As Murray observed at the end of the 1980s, the economic boom of that decade

> has been talked of as a national boom, resulting from the liberalization policies of national government. But quite apart from the strong expansion of the world economy, the boom as it appears in government statistics, needs to be loosened from its 'national' moorings in three ways.
>
> (Murray, 1989: 3)

The first of these, according to Murray, is that the boom itself was principally a boom *in* the south. It was, then, if we are to follow Murray's interpretation, 'a boom of the core', rather than a national dynamic of growth. Second, the pattern of growth was uneven in its impact across the south, bringing prosperity to some areas whilst by-passing other locations. And the third observation offered by Murray to disrupt the notion that the 1980s boom was national in character refers to the international nature of much of the growth that took place in the south east at that time.

In part, this last observation prefigures Smith's reference to the 'pull of Europe' – its capital and investments – as a factor consolidating the south east's importance and perhaps exceptionalism, but there is more to Murray's argument than that. In particular, Murray wishes to stress the neo-liberal strategies of liberalization and deregulation put in place by successive Conservative governments in the 1980s as a factor in attracting European and wider investments to the south east region (and to its neighbouring regions), especially those of finance and other key private services.

Such an account puts the workings of the 'free market', or rather the attractions of a lightly regulated market, at the heart of the south east's growth in the 1980s. There are many factors which can come to the fore in such a scenario, a number of which can be traced to a series of interventions by national governments. Among the best-known of these perhaps is the broad strategy of financial deregulation which, in the second half of the 1980s, led to a spectacular burst in credit and borrowing by individuals and companies alike. The enticing image of an economic boom as one fuelled by 'confetti money', as controls on lending – mortgage and credit-card lending in particular – were relaxed is one that still lingers among popular commentators. And indeed, such imagery is not entirely misplaced, especially as a clue to the kind of growth which took place in the late 1980s.

If, for instance, we take a short-term view of what happened across the south east in the 1980s, stressing events in the recent past, it is possible to weave a narrative of growth which presents this period as an exceptional economic moment, a one-off scenario. It runs something like this.

'Thatcherism' as a form of neo-liberal rhetoric may be regarded as a shorthand for the promotion of a market economy and all the individualistic, enterprising virtues that it is said to instil. Together with the 'Lawson boom' (Lawson was the Chancellor of the Exchequer under Margaret Thatcher for six and a half years until late 1989), Thatcherism can be seen as one part of a richer constellation of events which owe their significance to a combination of national and international forces at work across the south east in the 1980s. Nationally, with the lifting of exchange controls in 1979, a chain of events was set in place which would end with Lawson first fuelling a runaway boom in the late 1980s and then presiding awkwardly over its collapse (Smith, 1993). The removal of exchange controls, followed by a number of deregulatory measures, not only resulted in an immediate increase in overseas lending by UK banks, it also foreshadowed a general rise in financial lending. As is now well known, this, many have argued, led to a rapid growth in mortgage lending, credit-card debt and other forms of personal debt, which in turn stoked the booms in the housing and consumption markets in the late 1980s. Spiralling house prices and a consumer-led economic boom, financed in part through equity extraction from rising property values, gave the boom a heightened sense of sustained growth, especially in the south east.

Internationally, too, the 'Big Bang' reforms of 1986 which lifted restrictions on dealings in equities and bonds exposed the City of London to a more competitive world of finance which contributed in a variety of ways to a 'boom in the south'. The liberalization of financial markets not only created the conditions for a boom in investment in commercial property in the City, as finance houses upgraded or erected buildings replete with state-of-the-art communications technologies, it also led to a rapid rise in employment in the financial services sector. The impact of this job growth was not restricted to London; it occurred also, for instance, in the home counties, generating, in turn, demand for particular types of goods, services and housing, often on the basis of highly inflated salaries (Thrift and Leyshon, 1992).

Looked at in this way, the 'boom in the south', as an exceptional bundle of political and economic events, points towards a particular kind of growth; namely, one based upon speculative gains, corporate and personal debt and, more broadly, private sector services. On this account, the south east as a 'growth region' in the 1980s represented a quite different combination of forces from those surrounding the growth strategies adopted by, say, the Wilson administration in the 1960s towards the modernization of British industry. The stress then upon the modernization of manufacturing, harnessing the powers of science and technology to transform industry and create the broad conditions for social welfare, contrasts sharply with the emphasis placed by 'Thatcherism' upon liberalization, private services and an open stance towards the internationalization of the economy.

More significantly, whereas growth under the Wilson government in the 1960s sought its electoral base in the 'old industrial regions' of the north, the political success of the neo-liberal strategy was dependent upon the social and economic base of the south east (Massey, 1984; Jessop, Bennett and Bromley, 1990). As a region privileged by finance and private services and arguably the most open of all the regional economies, the south east was well placed to take advantage of what Thatcherism had to offer. In that sense, the 'success' of the south east was taken to be emblematic of the neo-liberal political project and was represented as such.

What was all the more remarkable though about this celebration of the south east was that it broke the longstanding consensus between the major political parties (formed in response to the 1930s' depression) on regional development and equality in favour of a principle of 'winners and losers'. If the north was unable to follow in the neo-liberal footsteps of the south, then it had only itself – its local politicians and entrepreneurs – to blame. Whilst obviously many in the north saw things differently, on the strength of the neo-liberal interpretation of events the dominance of the south east came to be represented *through* a particular kind of growth and *through* the new social groups in the region.

On this short-term view, therefore, the political and economic events which affected much of the country in the 1980s were the product of a particular neo-liberal moment, of which the south east was the main beneficiary. The success of the region in the 1980s may thus be traced to this exceptional political and economic moment. As *the* growth region, the dominance of the south east was reaffirmed.

This is only one narrative of growth, however. By adopting a different time-frame, one which stresses the long-term nature of the region's dominance in the UK, an alternative interpretation of what happened in the south east in the 1980s is possible. On this view, the dominance of the region in the 1980s was not altogether surprising: given the historical economic and social structure of the region, it would have been remarkable had it not experienced growth. In this narrative, which dates back to the 'exceptionalism' thesis first proposed by Perry Anderson in the early 1960s, the south east represents the hub, as it were, of a national economy in which the interests of the 'rentier' and landowner were placed above that of the manufacturer. The patrician stamp of the nation's ruling bloc and the predominant role of the City in the country's economic and political affairs were said to have had a distorted effect upon the shape of the British economy – with significant geographical consequences (see relatedly Ingham 1984, 1989; Barrett Brown, 1988).

There are a number of strands to this line of thought, not all of which are compatible, although broadly speaking the dominance of the south east is constantly represented as a feature. In the first place, Britain is regarded as 'exceptional' in comparison with its western counterparts because of its archaic polity: a fusion of landed and commercial interests which were successful in overriding and incorporating the rising interests of manufacturers at a time when the country was widely regarded as the 'workshop of the world' (Anderson, 1964). This coalition of interests was said to favour the City and commerce at the expense of manufacturing, with the result that, according to Lee (1986), Britain in the nineteenth century represented two relatively separate regional economies: one based on the accumulated wealth from trade and finance in London and the south east, and relatedly the landed wealth in the adjoining shires, and the other, less significant, economy in terms of size and wealth, based on mining and manufacturing in the north. In consequence, the south east was not only regarded as the significant base of the British economy, it was also considered to be the site of social and political power more broadly. Dominance was an effect also of a marked geography of class.

This state of affairs has in turn been extended into a form of cultural dominance, with the south east at the centre of a network of influential institutions and cultural practices. The argument has been expressed most forcefully by Weiner (1981), as mentioned earlier, in terms of an 'anti-industrial spirit' which has long pervaded English culture and privileged a range of more 'gentlemanly' values which have more comfortably been employed in the professions, commerce and a lifestyle of (synthetic) ruralism. Others, such as Rubenstein (1994), have emphasized the reverse effects: the 'gentlemanly' character of British capitalism as the direct product of an economy which even in the nineteenth century was commercially and financially orientated. What emerges from all accounts is the key importance of London and the south east within the networks of privilege – between financiers and the landed gentry, between the public and private professions, Oxbridge, the public schools and the Anglican Church – as a constant feature of the cultural landscape.

Putting all this together – the cultural capital of customary elites in the south east, the predominance of the City and commerce in the national economy, and the asserted international, indeed cosmopolitan, character of the southern middle class – the long term dominance of the region in the UK is not altogether surprising. In this narrative, then, the political comings and goings of 'Wilsonism' or 'Thatcherism' are of little consequence compared to the continuities of south east civil society. The burst of growth in the 1980s which benefited the south east was in this longer view quite unexceptional. What would have been astonishing is a scenario in which, given the cumulative advantages of the region, the pattern of growth had by-passed the south east altogether.

Recalling our earlier discussion, in taking both a short- and a long-term view of the fortunes of the south east, we have tried to draw attention to some of the ways in which the dominance of the south east has been reproduced. Central to both accounts, whether the dominance of the region is understood in terms of recent growth or assumed as part of its long historical inheritance, is that that dominance is as much about the make-up of people, the social relations of the region – about class and style – as it is about identifying blocs of economic or political power. As we shall see, this social inheritance is both essential and contested. The two accounts when placed side by side also provide some understanding of the discursive context in which neo-liberal attempts to represent the south east as *the* growth region in the 1980s took place. The region was already the central figure in a regional discourse of dominance; turning that into a story of dominance through growth was not at odds with earlier positionings. Notions of the south east as a core or lead region were, after all, part of earlier representations within cultural and political as well as economic discourses.

What are less clear in accounts which juxtapose short- and long-term narratives, however, are the different histories, break points, and qualitative shifts in the make-up of the various characteristics of the south east which led to its status as a growth region in the 1980s. Regions are, after all, constituted by their place within a wider constellation of forces and events, some of which may come out of long-running shifts in the structure of a region or society more generally, whilst others take their shape from a particular historical moment. In our view, the south east as a growth region was constituted by the coming together at a particular moment of a set of dynamics of growth. It was these which provided the propulsive force for the expansion which the region experienced. Moreover, these different dynamics of growth which made up the south east in the 1980s can each be said to possess their own history, have different starting-points and operate according to their own rhythms – as well as laying down their own geographies. Unpacking these differences and analysing their development is of considerable interest to us here, and so too is an understanding of how the different components came together in the 1980s to produce not only a distinct geography of the region but also a qualitatively different form of growth from that of previous periods. Such complexity is interesting because what it implies is that the south east in the 1980s was the product of the intersection of particular social relations, a particular – grounded yet still momentary – formation of social relations in a wider spatio-temporal complexity. It is therefore to this issue of growth that we now turn.

Components of growth

One of the central mechanisms of growth which has figured constantly in accounts of the south east's dominance is that of *finance*. As we have seen, there are long-term reasons why finance has played a significant role in the economy of the south east and the UK more generally. The simple fact of the presence of the City of London and its accompanying commercial activities has produced an accretion of economic and cultural capital over time. More than that, the outward, international orientation of the City's business practices since at least the nineteenth century has long given the impression of a financial space which has pulled itself away from the rest of the nation's economy. On this view, then, the deregulation of finance in the 1980s – the abolition of exchange controls, the reform of the Stock Exchange, and so forth – merely represented an extension of this scenario. The City of London and its financial activities remained a space apart and a source of dynamism which appeared to have few consequences, other than negative ones, for the rest of the national economy.

While the long-term character of London and the south east's financial and commercial dominance in the national economy is beyond reasonable doubt, there have also been qualitative shifts in the growth of financial services in the region since the 1960s. Perhaps the most significant shift is that outlined by Pryke (1991), who attempted to capture the changes under the title of 'an international city going global'. By this he meant that until recently the UK's financial sector had been locked into the international framework of the Sterling Area and Empire, but with the development of new financial markets, new financial instruments and a series of space-shrinking technologies, the very scale and tempo of British financial practices has been disrupted. Finance as an axis of growth in the 1980s, in this sense, then, was about the new type of financial activities performed in London and the south east which tied it into the flows of global capital and money markets. Not all of these activities had their starting-point in the 1980s, however.

In the first place, there was the development of the Euromarkets in the 1970s which led to a spectacular increase in the internationalization of bank lending, of which London (and the overseas banks attracted to it by the massive flows of funds) acts as the main 'clearing house' (Coakley and Harris, 1983; Coakley 1984). Whilst the flow of funds through London reinforced its historical entrepôt role and the City's divorce from the rest of the UK economy, the dominant role of overseas financial institutions in international lending signalled a break with the 'gentlemanly' practices and rhythms of the City's past. In the foreign exchange market too, largely because of developments in Eurocurrency business, overseas institutions are the main players in London's dominance of this field of finance (Coakley, 1992). The same is true of the Eurobond market, whose activities are linked to a range of new financial instruments such as equity warrants and swaps. In each of these markets, therefore, change rather than continuity is the hallmark, with British financial practices having to adapt to global rather than Sterling's days of finance.

In this context, the deregulation of finance in the 1980s, the 'Big Bang' reforms that we have already spoken about and the Financial Services Act of 1986 in particular, can be interpreted as a direct attempt by government to reorientate the City's traditional institutions to global markets (Leyshon, Thrift and Daniels, 1987; Jessop and Stones, 1992). A more global City had, however, emerged from the myopia of the 'gentlemanly' institutions of finance before Thatcherism took hold. In combination with the Thatcherite reforms, such changes produced the boom in financial services in the 1980s which benefited London and the south east directly – in terms of job numbers, foreign earnings and massive investment.

In employment terms, for example, regional growth in financial services was dramatic. In the second half of the 1980s, between 1984 and 1990, 84,000 new jobs were added to London's financial and commercial complex (Rajan, 1990) and in the region as a whole financial services employment grew by 88 per cent over the decade (Leyshon and Thrift, 1993). In investment terms, too, the rise of a transaction-driven, securitized financial system prompted financial institutions to undergo extensive equipment refits to obtain the information capacity, dealing and settlement systems necessary to compete within the new global financial markets (Leyshon, Thrift and Daniels, 1987). This in turn led to a demand for new buildings to accommodate the new communications technologies and to further economic multipliers.

But, our concern here is with the *structural significance* of finance as a growth mechanism. Whilst the spin-offs from growth in the financial services sector, in terms of the demand for commercial property or high-technology systems, were significant, as we shall see later, and the job numbers created indirectly (including those way beyond the standard region) were an important component of that growth, it is the nature of the financial activities which took place in the region and the global networks of relationships in which they operate which made finance a growth mechanism. Some of these activities and the institutional practices which support them spring from the 1980s and the neo-liberal framework put in place over that period. Most, however, pre-date Thatcherism by at least a decade, whilst the economic power of the City's institutions, although disrupted by the shift in global orientation and strong competition from other financial centres, represents a long-term underpinning of London and the south east's economy.

A similar kind of analysis can be offered for another dynamic of growth which held a central position in accounts of the south east's boom in the 1980s: namely, that of *consumption and debt*. By looking at the nature of consumption in the south east, it is possible to demonstrate through the work of Lee (1984, 1986), Rubenstein (1977, 1981, 1987) and others, the long-term affluence of the region in terms of wealth and income. Rubenstein's initial study of wealth, occupation and geography among the Victorian middle classes, for example, points towards the concentration of wealth in the south east in the nineteenth century, much of it derived from finance and commerce, and his later work suggests a highly skewed distribution of income and personal wealth in favour of London and the south east. Lee's investigation of regional growth patterns since the Victorian period arrives at similar conclusions, extending the analysis to the best part of the twentieth century. Such regional differences in income levels have persisted, and indeed became exacerbated in the 1980s.

It would be misleading, however, to stress the persistence of such a *pattern* of inequality at the expense of the *processes* which, at different times, have produced it. In looking at the nature of consumption in the contemporary period, we need to distinguish between the accumulated wealth and income derived from finance and trade over the long term and the more recent events – referred to earlier – which led to the spectacular burst in credit and borrowing in the region. More importantly, in considering consumption as a growth mechanism we are not concerned with each and every purchase that took place in the region throughout the 1980s, but rather with those aspects of consumption which can be considered as directly propulsive of growth, such as housing and other forms of property, or purchases within high-value retail markets.

Clearly, the relaxation of restrictions on financial lending played a significant part in the consumption-led boom of the 1980s. Consumption debts among households rose dramatically in the 1980s across the south east, the majority of them taking the form of mortgage debt, followed by more traditional forms of consumer borrowing such as the advancement of loans, and credit-card debt (Smith, 1993). With the increase in competition from banks in the mortgage market, coupled with easier access to funds, net mortgage lending grew at an astonishing rate for much of the 1980s. At the height of the housing boom in 1988, lending was more than four times that of the 1980 figure (Hamnett and Seavers, 1994a). Much of this lending, however, found its way not into bricks and mortar, but into general consumption, especially in the south of England. Equity extraction in the broad south east, therefore, provided a considerable boost to consumption in the region as homeowners unlocked their capital on the expectation of rising house prices further boosting their personal wealth.

Higher-income households in the south east, most of which are likely to be owner occupiers too, were also able to boost their consumption through the use of loans and credit cards. Financial institutions, from the retail banks to the building societies and some of the larger retail stores, made significant inroads into credit money markets in the 1980s, contributing to the overall rise in personal debt. Whilst it is difficult to put a precise figure on the extent to which credit-card debt fuelled the boom in the south east, the strong correlation between credit card use and high-income households is an indication of its role in supporting the boom. With the largest concentration of high-income households in the country, the impact of rising consumption levels and the associated fall in the savings ratio made the region the most indebted part of the UK in the 1980s. In this context, Lawson's budget tax cuts in the late 1980s really did give rise to the impression that consumer spending was being fuelled by 'confetti money'.

Affluence in the region, therefore, especially among the new enterprising professional middle classes who saw their income levels rise appreciably in the 1980s, contributed substantially to the pattern of growth across the south east. The processes by which the consumption-led boom was generated, however, owed as much to the accumulated wealth of the region's households whose personal wealth formation was built up over time from finance and commerce, as it did to the political and economic events of the late 1970s and 1980s (see Savage *et al.*, 1992; Thrift, 1996). The different histories of the social groups involved came together to make consumption a growth mechanism in the 1980s.

Related to this, as noted earlier, was the investment which did make its way into residential property which, along with the overall investment in the built environment in the region, made up a broader property dynamic in the 1980s. Once again, the idea of property as a growth mechanism has less to do with the pattern of its distributional benefits, for example in terms of rising house prices or the number of jobs generated in the construction or exchange industries, and rather more to do with the nature of the region's property markets and how they changed over the 1980s.

In residential property, for instance, the market underwent a rapid expansion in both its regional share of new private dwellings and turnover in the second-hand market. Much of this can be traced to the liberalization of housing finance, and to the extraordinary set of events which led to a limitation on mortgage tax relief by Lawson. This tax change produced a dramatic rise in house prices, especially in London and the south east, which saw the housing 'bubble' burst, first in the region and then elsewhere. In one sense, there was little that was new in the cycle of boom and slump in the region's housing market – except, that is, for the abruptness in the rise and fall of house prices. While the pattern was a familiar one in the post-war period, however, the rhythm was not, and this may in part be traced to the Thatcherite emphasis upon allowing markets to find their 'own level' in a recession (Hamnett and Seavers, 1994a).

Booms and slumps have also been a recurring feature of the commercial property market in the south east over the post-war period, but few have been as dramatic as those witnessed in the 1980s. Around two-thirds of new office building took place in the region, with an equivalent share in overall investment. Moreover, because office rents were much higher in the south east, the region accounted for around four-fifths of office income in the mid-1980s. Indeed, the high share of new investment and the size of rental incomes are a reflection of the historically high cost of land and buildings in the region (Hamnett, 1985).

Commercial property markets in London and the south east have frequently experienced growth rates higher than the rest of the country owing to the demand for office space generated by the region's longstanding international role in finance and trade. From the 1970s on, however, as the nature of London's financial markets changed in orientation, so too did the regional property market, with the City of London in particular acting as a funnel for overseas funds. Aside from developments in Docklands and Canary Wharf, foreign investors were active players in quality buildings situated in London's Square Mile. That the market for commercial property in the region collapsed in the early 1990s, witnessing the withdrawal of a number of overseas investors, most notably the Japanese, does not imply a return to more parochial markets, however. The increasingly global, and essentially the European, nature of the region's property markets remain and it is this characteristic which makes property a structural mechanism for growth in the south east (Pryke, 1995).

In each of the above dynamics we have been keen to distinguish those aspects which generate economic growth – that is, the propulsive processes and activities involved – from those aspects which distribute effects. Only certain aspects of finance, consumption and property were considered to be generative of growth, whilst others represent part of the growth-transmission process. Although it is difficult to arrive at a hard-and-fast distinction between the two sides in all instances, given the interconnected nature of the growth dynamics, the distinction does draw attention to a particular way of conceptualizing growth. In the case of the next growth mechanism to be outlined, one whose history is of recent origin, that of high technology, the distinction is perhaps more easily grasped.

Not all industries within the high-technology sector may be considered propulsive. In fact, much so-called high-technology activity involves relatively routine production tasks, such as assembly, rather than those of innovation. Many of those aspects of high-tech activity which act as the bearers of innovation are to be found in industries like telecommunications, computing services and the independent research and development sector. And what is particularly noticeable about many high-tech industries is their concentration in parts of the south east. Over half the total employment in the above three industries can be traced to the standard south east region, a quarter of it in London alone. In computer services, for example, around two-thirds of those employed in the sector at the end of the 1980s could be found in the south east (Cooke *et al.*, 1992). In contrast, the more routine sides of high-tech work were more widely dispersed throughout the country, with different regions tending to specialize in one or more high-tech manufacturing activity (Begg and Cameron, 1988).

The nature of high-tech work in London and the south east is therefore rather different from that undertaken elsewhere in the UK and this has produced a distinctive concentration of scientists and qualified technicians in the region. The reasons for this geographical concentration are multiple (Massey, Quintas and Wield, 1992). It is, in part, a product of the state's involvement in the location of government research establishments and the outcome of Ministry of Defence expenditure, with recent changes in the defence industry reinforcing a south east bias (Lovering, 1991; Morgan and Sayer, 1988). The long-term presence of scientific elites in and around London and the south east, together with a history of product and process innovation in the region, have arguably also played a role in attracting the more propulsive side of high-tech industry. Moreover, this long-term presence can itself be linked into that more ancient geography of the British class structure to which we have already referred. As, over the decades, the contribution of 'science' to commercial production has become separated-out as a distinct activity – research and development – so those who found employment in these activities joined the flight from manual production (Massey, 1995b). A more recent development, moreover, is that such groups of highly qualified, professional and technical workers are now increasingly part of global labour markets which take in London and the south east at the expense of other regions (Fielding, 1992). Once again, the particularity of a dynamic of growth in the south east in the 1980s was the product of a multiplicity of temporalities. It is on these grounds that it is possible to talk about high technology as an axis of growth in the region throughout the 1980s and indeed into the 1990s.

The final dynamic to be considered here is of a rather different order from those mentioned above. Rather than a dynamic or growth mechanism, the interventions by a succession of neo-liberal governments in the UK from 1979 onwards helped to shape the nature of growth across the south east region. Indeed, their interaction is part of what the south east is today. We can put this another way.

On a number of occasions we have referred to government actions which have since the late 1970s been conceived broadly along neo-liberal lines. These would include the whole panoply of measures which fall under the general heading of 'privatization': namely, those of financial deregulation and the liberalization of money markets, the sale of nationalized industries and the expansion of share ownership, the sale of public housing and the promotion of homeownership, as well as the break-up of the NHS and the encouragement of private health provision. They would also include changes in the tax regime, both the reduction in the rates of income tax and the adjustments to mortgage tax relief. Taken as a whole it can be argued that the outcome of these policies disproportionately benefited households in the south east. On their own, individual policies may not have been directly propulsive for the south east, but the net overall change gave a boost to growth in the region by broadly re-defining the structure of a wide range of markets .

The impact of financial deregulation upon the growth of private services in the region, especially financial and commercial services, has been mentioned, along with the boost to housing and property markets more generally, and the increase in household and consumer debt. Figures for other measures, such as the distribution of share ownership or the impact of the tax 'reforms' also show households in the south east, and high earners in particular, reaping greater benefits than those in other regions (Hamnett, 1994).

In terms of direct government expenditure too, the south east region in the 1980s was the recipient of public funds which amounted to what many have referred to as 'counter-regional subsidies'. From the massive public expenditure poured into the Channel Tunnel and the Docklands projects to the public funding of road and airport expansion schemes (Stansted in particular), government investment decisions effectively boosted development in the south east at the expense of other regions. With the longstanding concentration of civil service functions and jobs, notwithstanding limited decentralization trends, and the benefit of lucrative MoD defence procurement contracts in the region, the south east benefited from government actions in a way that no other region did.

It is important to note again, however, that not all of this regional bias is of recent origin. The presence of the civil service machine in the south east is hardly new; indeed the fact of its being in the south east is, as Smith observed earlier, one of the reasons for the very dominance of the region. None the less, in the 1980s, this dominance was not achieved on the basis of such institutions alone, however strategic their role may be in the country at large. In the 1980s, the actual political decisions taken by government, as opposed to the physical presence of government, acted as a further propulsive element. And, moreover, one which helped to shape the region along the lines of a 'free-market' form of growth.

Looking back on the 1980s, and the post-war period as a whole, with the benefit of hindsight it is easier to see how successive government actions contributed to different forms of growth and development. As Leys (1985) argued at the height of the Thatcherite challenge, the endorsement of 'privatization' and the free play of market forces represented a conscious strategy of destruction of what had gone before. The practice of planning, the modernization of British manufacturing through technological change and the institutional transformation required by the social-democratic platform of 'Wilsonism' in the 1960s, were of little relevance – indeed, were seen as hindrances – to a political strategy and ideology which placed 'the market' at the core of national progress.

In the south east, however, it was as much the attack upon what had avoided or resisted modernization – the unmerited privilege of the established middle class, the customary elites and 'gentlemanly' practices of the home counties – which left its mark upon the regional social structure. Contemporary neo-liberalism and its market doctrine represented an attempt to undermine both the legitimacy of corporate social democracy and the practices and expectations of an older, 'gentlemanly' capitalism inscribed in the south east. If anything, it was this older network of privilege in the region which experienced much of the force of the neo-liberal attack in the 1980s and whose cultural and political dominance was shaken as a direct result. If the south east was now *the* growth region, it was so on the basis of a form of growth which clashed with many of the cultural and economic values of 'established' civil society in the region.

We shall take a closer look at the nature of this clash shortly, especially in relation to the new lines of division and advantage which emerged. What is important to note here, however, is that the clash itself, although orchestrated by the Thatcherite project, was not reducible to it. Each of the components of growth that we have considered here contributed towards a remaking of social relations in the south east in the 1980s. Individually, taken on a one-by-one basis as we have done here, there is nothing particularly remarkable about their social impact. In combination, however, as a series of dynamics, each with their own rhythm of development and relationship to certain enduring elements of the south east, they gave rise to a highly particular moment, and a particular form of growth. In the 1980s, the different components of growth came together in the region in such a way as to produce a qualitatively different form of growth from that of previous periods. Under the broad political project of neo-liberalism, the axes of growth which had a structural presence in the south east were galvanized in a particular direction: along new lines of inequality and advantage, which benefited new social groups at the expense of others, especially at the expense of those at the bottom end of labour markets and in the poorest of households (see Mohan, 1995a; Townsend, Corrigan and Kowerzik, 1987). Not only was the economic gap widened between the south east and the rest of the country, but new dividing-lines were opened up within the region itself.

Forms of growth

That growth may take particular forms, that the benefits of growth will flow to different groups at different times depending on the form of growth, will come as no surprise to those excluded by the events of the 1980s. As Jessop *et al.* (1989, 1990) have stressed, the Thatcherite project involved a political shift in emphasis from a 'one nation' approach which held out the prospect of a general improvement in living standards for all, to a 'two nation' strategy in which

> its main social basis comprises the beneficiaries of the entrepreneurial society and popular capitalism – above all the richest groups in Britain or as Rentoul (1987) labels them, the 'have lots'. . . . Numerically the principal beneficiaries are the new service class especially its private sector professional and managerial strata engaged in finance, business services and semi-skilled private-sector manual workers in core parts of growth industries. Both groups also have a strong (but not exclusive) regional base in the south notably London and the growth triangle bounded by Cambridge, Bristol and Southampton.
>
> (Jessop *et al.*, 1990: 96)

If the 'have lots' in the south, principally the new middle class in the private services sector, were favoured by the Thatcherite project, those that missed out or were directly excluded by the 1980s boom were not simply outside of the growth areas. In the 'heartland' of the neo-liberal project, those dependent on welfare benefits, in particular single-parent households and pensioners, as well as the least-skilled or workless pushed into insecure, low-paid employment, found themselves on the 'downside' of growth (Allen and Henry, 1995; Massey and Allen, 1995; Mohan, 1995a). Although the Thatcherite project arguably set out intentionally to divide the nation, both socially and geographically, it also sought to transform the country as a whole into an enterprising, competitive economy. To achieve this, however, those people and places which stood in the way of or represented an obstacle to this transformation took the brunt of the neo-liberal strategy. The reality, therefore, as Jessop *et al.* described it, was a 'two nation' politics, with the new working poor and those dependent on state benefits forced into the category of 'have nots'. What characterized the Thatcher years was this widening gulf between the 'haves' and the 'have nots' and the marginalization of those groups which did not mesh with the needs of this particular form of increased economic efficiency and enterprise (Jessop, Bennett and Bromley, 1990).

However, while the fact of polarization is now widely recognized, the complexity of its form and nature is less often addressed. There are a number of aspects to this. First, and as we shall explore in much more detail through this book, it had its own, complicated, geography. While the south east soared away from the rest of the country in terms of average per capita income, the region was also the one with the greatest degree of inequality within it. Moreover, second, this growth in income inequality – both inter-regionally and intra-regionally – was more than anything based on the concentration of very high salaries in one particular group of the population – the top decile of male (and only male) earners in the south east (Massey, 1988a). It was, then, a highly socially and geographically specific polarization. Moreover, third, at the other end of the scale from these high-earning men, two different causes of poverty (and thus of this aspect of polarization) must be distinguished. On the one hand, there was the poverty which resulted from straightforward exclusion from the dynamics of growth: the parts which growth did not reach. On the other hand, there was the structural inequality within the nature of the growth itself (the contrast between financiers in the City and those in the newly 'flexibilized' services who clean their offices, or between top management and casualized lower-echelons in privatized utilities come to mind). Each of these forms of poverty (through exclusion and through impoverished inclusion) has, again, its distinct geography, as we shall see. But the major point to underscore here is that inequality was *structurally inherent* to this form of growth. It was not (only) that the growth did not spread widely enough, it was also that inequality was exacerbated in the very process of growth itself.

The neo-liberal decade is, of course, now (in)famous for this characteristic. Both between countries – developed and less developed – and within them, it was divisive.

André Gorz in his *Critique of Economic Reason* (1989) has drawn attention to one potential mechanism, within industrial countries, of this widening gulf between the 'haves' and 'have nots'. He postulates a sharpening divide between a group of workers able to monopolize the well-paid, professional jobs and another group of workers who are effectively constrained to 'service' these professional classes, in terms of their domestic and personal service needs. On this view, recent changes in the nature of work and the economy have operated to the advantage of those already in secure, well-paid, middle-class employment. These people are 'work-rich'; they have professional careers which increasingly leave them little time for much else besides work and certainly limited time to 'service' themselves. On the other side of the divide are the 'work-poor', those who have been marginalized or excluded by the direction of economic change in the advanced economies – those who have lost their jobs through the decline of the Fordist industries or through technological shifts. Faced with limited work opportunities, this group of people have few employment options, in Gorz's view, but to meet at low cost the personal service demands of the 'work-rich' middle classes. As a type of private service 'elite', the professional classes are quite simply 'able to purchase time more cheaply than they can sell it personally' (Gorz, 1989: 5). Gorz speaks about the emergence of a 'servile' class organized by private service firms on an insecure, casualized basis and then hired out to meet the cleaning and childcare needs, for example of 'work-rich', professional households.

Gorz himself provides little in the way of empirical evidence for this form of growth. However, in the UK, the study by Gregson and Lowe (1994) on the resurgence of demand for waged domestic labour by the middle classes in the 1980s and early 1990s did provide some form of qualified support for his assertions. Among professional, middle-class households, they documented a crisis of domestic labour, especially around childcare, although its extent was limited to a sizeable minority of such households. Equally, they found that the 'crisis' in getting the domestic labour done in middle-class households was not as straightforward as Gorz would have us believe, often reflecting a desire rather than a need to 'buy in' a particular service like cleaning. Moreover, the story has a strong gender component, involving the rigidities of certain forms of masculinity and crises of femininities pulling in different directions, which Gorz entirely fails to address. While women increasingly found employment outside the home, men failed to increase their contribution to domestic labour and low-wage female labour had to be brought in to compensate for their under-performance. While the evidence in the UK for Gorz's type of divided society is therefore limited, it is interesting to note that this later study did highlight one factor significant to us here; namely, that the resurgence in demand for waged domestic labour was concentrated in London and the south east region in the 1980s.

Perhaps the most significant weakness of Gorz's broad account of polarization, aside from its neglect of gender, is its abstraction from politics, and national politics in particular. The nature of the divide between 'haves' and 'have nots' and its extent cannot be divorced from the political regimes of different countries. The liberal, Anglo-Saxon model – as Esping-Anderson (1993) and others have argued – gave social polarization a shape that was quite distinct from that under social-democratic, welfare regimes. In the Nordic countries, for example, the pattern of Fordist job loss was met by a government-backed expansion of welfare services and secure employment, which has done much to counter the growth of low-paid, insecure jobs at the bottom end of the private services sector. In countries like the UK, however, the 'mass of bad jobs', according to Esping-Anderson, has grown on the basis of low pay and minimal security. In a de-regulated labour market, in the sectors of the economy which operate on a hire-and-fire basis, the level of wages is likely to be pushed down and the precariousness of work raised in profile (see Allen and Henry, 1996). The increasing inequality of the 1980s, then, was a complex phenomenon. And rather than a simple economic gulf between 'work-rich' and 'work-poor', the precise nature and extent of the divide rests upon the form of political intervention and the nature of the social settlement entailed. It is in this sense that it is possible to talk about a political project conducted along neo-liberal lines attempting to transform social relations in the UK in the 1980s.

Moreover, that political project gave shape to a form of growth which besides being inherently unequal also cut across existing lines of privilege and advantage. To describe the neo-liberal project only in terms of 'haves' and 'have nots' is to miss the nuances of a 'two nation' politics which sought not only to confront those 'who would not help themselves', but also to challenge traditional 'one nation' Conservatism and its assumed advantages.

Gamble, in his *The Free Economy and the Strong State* (1988), offers a convincing account of the nature of that challenge and the contradictory way in which elements of the established middle class both benefited and lost out from the Thatcherite project. Under the 'one nation' politics of traditional Conservatism, the established middle class, in particular the propertied and professional classes, were the upholders of many of the country's major institutions – the Church, the Monarchy, the Civil Service, the BBC, the 'old' professions around law and medicine, and the like. All, as Gamble points out, were once regarded as bastions of the status quo by Conservatives across the country. Now, however, such institutions and the classes which stood behind them were considered by Thatcherites to be anti-enterprise, and therefore blocking the path to radical change. Under a neo-liberal strategy the culture of these establishment institutions was openly challenged, although in ways that produced quite contradictory outcomes for many of the propertied and professional classes.

The 'gentlemanly' traditions of doing business in the City of London were challenged as outmoded in today's global financial economy, yet the City itself and indeed many of its traditional players prospered under more open, de-regulated financial markets. The anti-enterprise culture of the propertied classes, many of whom derived their income from rents and dividends on investments rather than from work, was also derided by Thatcherites, yet the rentier elements prospered in the 1990s as investments abroad were encouraged. The paternalistic, rather aloof practices of service associated with the 'old' professions and the establishment institutions were scorned by neo-liberal conservatives, but the majority of professionals did rather well materially in the 1980s. Indeed, it could be argued that what the Thatcherite project directly challenged was not so much the economic dominance of this middle-class grouping as its cultural and political dominance. While the policies of neo-liberal governments did not set out to favour the interests of the propertied and old professional classes, their impact, as Gamble notes, frequently worked to their economic advantage. The home-based economic interests of much of the UK's manufacturing industry, not for the first time, fared less well.

The challenge, then, to the customary elites was effectively a challenge to their political dominance – as a kind of fetter upon radical change – and a challenge to the assumed authority of their class culture. In other words, it was a challenge to the identity of this middle-class grouping which attempted to speak on behalf of the nation. As such, Thatcherism locked swords with a gentlemanly style of 'Englishness' which has its roots in the Victorian era and Anderson's 'exceptionalism' thesis. More to the point, it took issue with the 'Englishness' of the home counties and the metropolitan elites. For, as Hall (1992) among others has pointed out, there is no single national identity – no one version of 'Englishness' or 'Britishness' – but rather 'competing national identities which jostle with each other in a struggle for "dominance"' (ibid.: 206). 'Englishness', she argues, is an historically contingent construction: a white culture which, in the mid-nineteenth century, was represented through a particular middle class culture and a particular masculine coding. It was also, we should add, a *placed* culture. It was a culture which, although not unique to the south east of England, only made sense in relation to the geography of (parts of) London, the home counties and, indeed, the Empire.

In the post-war period, the contours of this hegemonic culture, its characteristics of class, of gender, of ethnicity and of geography have been ever present. Wilson's modernization strategy was an attack on the amateurism, the assumed privilege and the accumulated traditions associated with this elitist premodern culture. Like Thatcherism, Wilson's cultural attack failed to shift this dominant cultural bloc, but where the neo-liberal project did make ground was in promoting groups and places which defined themselves in relation to, or in contrast to, the predominant culture of what may now be referred to as the white ROSE (the Rest of the South East, after London has been excluded).

Referring to the dominant culture of 'the white ROSE' as an identity with which Thatcherism took issue is not to suggest, however, some kind of simple concern with skin colour. Rather, the white Englishness of the ROSE is a sign and symbol of a more complex social construction: one that is *gendered* white and which draws its whiteness from the networks of privilege which make up the established southern middle *class*. It is a gentlemanly whiteness, an Oxbridge whiteness, a City whiteness, and a landscaped whiteness which is to be found in the more 'scenic' parts of the south east. Thatcherism, as we have suggested, presented a challenge to this cultural identity. And in the south east of England, this took the form of new social groups defining themselves in relation to this dominant culture. As Jessop has highlighted, foremost amongst these were members of the new professional and managerial middle classes, but the decade

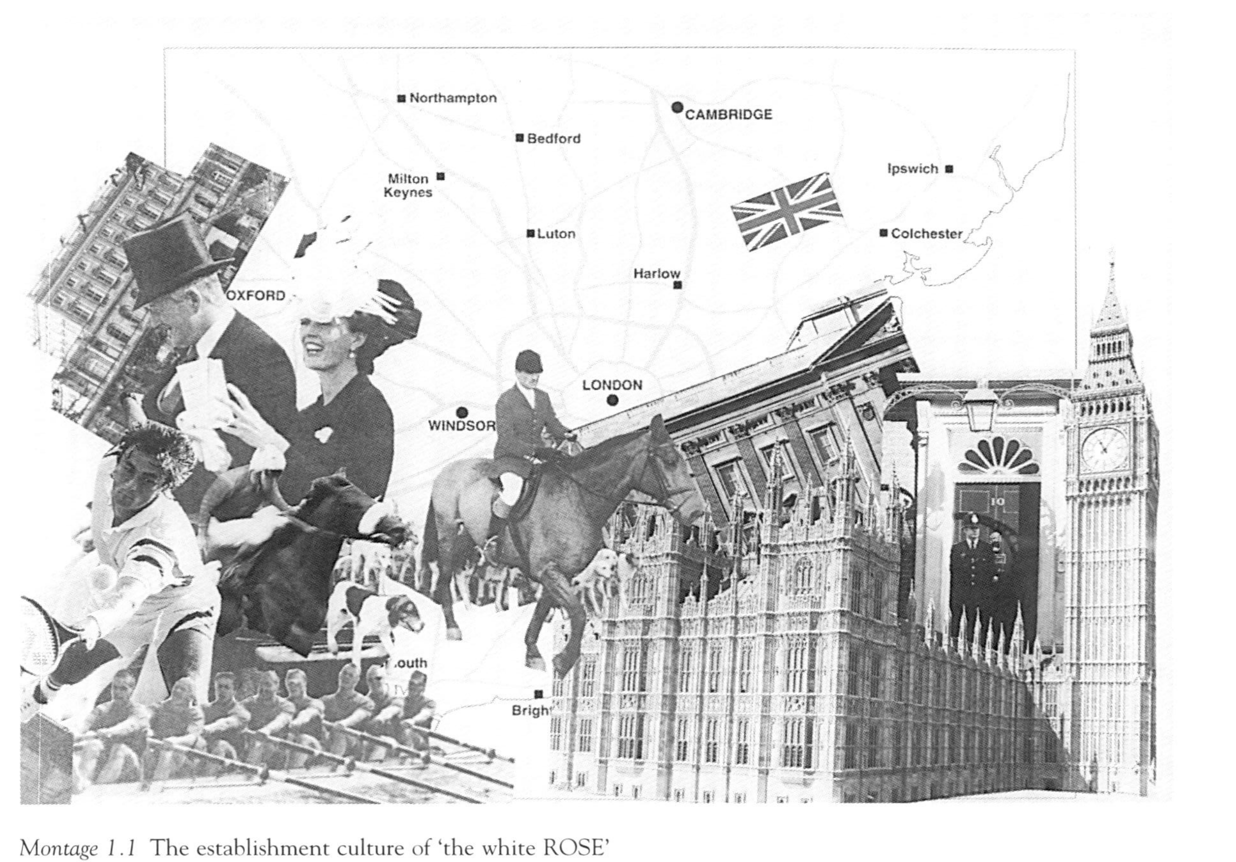

Montage 1.1 The establishment culture of 'the white ROSE'

also saw the rise of an 'Essex person' renowned for a lack of suitable reverence for genteel manners and values.

In the 'growth region' of the 1980s these emerging identities were constructed in relation to the dominant culture of the white ROSE and in relation to other adjacent groupings. If the south east region lay at the core of the Thatcherite project, its dominance in the 1980s was represented through a celebration of a set of middle-class identities which challenged – but did not supersede – the older forms of white English masculinity of 'one nation' politics. This too was a white culture, but the whiteness in this instance signified a set of class characteristics which represented a new articulation of social elements, both adapted from the old and celebratory of the new. Members of the new middle class took on and modified the symbols of the past, in part changing their meanings and in part drawing on them to confirm their legitimacy. The rise of four-wheel-drive vehicles (even for urban transport), the ubiquity of green waxed jackets and the rise in demand for country mansions all testified to the continued power of the rural (or at any rate, the Home Counties) idyll as representation of success in England (Thrift, 1989). But this was combined with a particular form of entrepreneuralism which projected a rather different message to the nation from that of the old networks of privilege and domination.

The south east had long lain at the heart of growth, but it was to be significantly reworked in the process of the reproduction of its dominance.

2

WHERE IS THE SOUTH EAST?

But where is this south east which was both created and reworked by 1980s' Thatcherism? We have been talking about the place for many pages now, and still have not defined it. Geographers, before they do anything else, surely should define their spatial framework. And we shall indeed do so presently. But – we want to argue – we could not have done so before we started the research. The previous chapter argued (and subsequent chapters will elaborate) that the Thatcherite project created a new south east, or at least that it produced something called the south east in a new form. It did so both discursively and materially. What this chapter will argue is that it did so also geographically. The last chapter examined a particular temporal constellation; the present one will examine the spatiality of that constellation. The growth of the 1980s created a new space–time: 'the south east of the 1980s'. In order fully to apprehend the meaning of this, however, we need to move away from previous approaches to regional definition and think about the issue in a wholly new way.

Defining the region

Part of the specificity of the neo-liberal form of growth in the 1980s was its geography. Its spatially concentrated nature, its division of the country into past and future, north and south, was one of its most remarked-upon characteristics. The north–south divide was a topic of concern within the media, in popular discourse, and within the ranks of the governing political party. A 'regional study of growth', therefore, in a broad sense, fell naturally into place as a focus for research.

It soon became clear, however, that a more precise definition of the region in question was by no means a simple matter. The question 'What and where is the south east?' became the central focus of our early research. The south east which was roughly encompassed by the discourses of national geographical divide in the 1980s was in fact one which stretched beyond the boundaries of the Standard Region but where, within those spatial reaches, were also areas untouched, or relatively so, by our specified mechanisms of growth. This is a part of the country often considered to be focused on London (though this is a characterization we shall question) and which in fact contains within it enormous variety. It includes the rolling acres of 'the Home Counties' – a countryside, still affluent though changing, which has been made emblematic of the English nation; and it also includes, most especially in parts of London, some of the highest concentrations nationally of ethnic minority populations. It includes the manicured countryside and the symbolic gathering places of the English upper class (Henley, Wimbledon, Lords, Ascot) and the mean streets and windy wastes of the East End of London and the Thames estuary suffering the long decline of manufacturing and dockwork. Here are both preserved country villages, and towns which grew up in the twentieth century on 'Fordist' manufacturing industry – Luton, Slough, Basildon, Welwyn Garden City and Dagenham. This 'region' is the point of arrival for most international tourists, and many venture no further. Apart from London, in this quadrant of England are gathered many of the classic sites of 'historic Britain' – Oxford, Cambridge, Windsor, Canterbury. And not just for tourism, but for flows of international migration and foreign investment too this is the part of the country which functions more than any other as the key hinge in the relationship between the UK and the rest of the world. It is, as noted earlier, above all the location of the prime national seats of power: political,

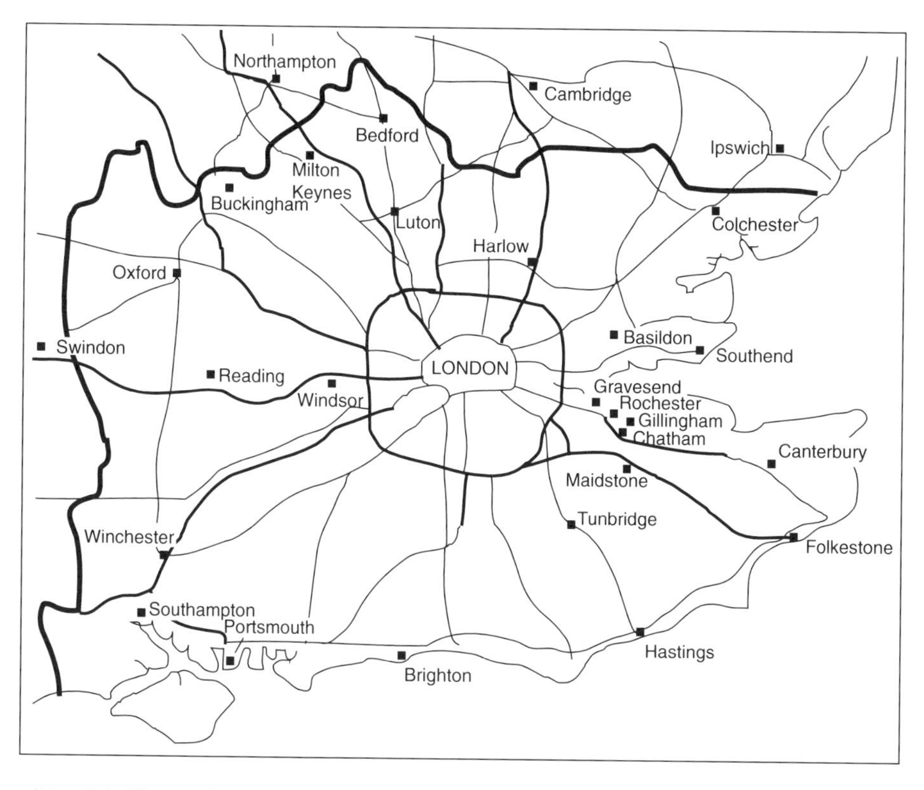

Map 2.1 The south east Standard Region

economic, cultural, military. It was going to be impossible to capture all this in one/single/'correct' definition of 'the region'.

We begin from the proposition that 'regions' (more generally, 'places') only take shape in particular contexts and from specific perspectives. There will always be multiple, coexisting, characterizations of particular spaces/places. The different social groups within a place may have different, even opposed and contested, readings of its character. Wider discourses – political, cultural or economic – may yield yet other identities. There is, we would argue in general terms, no 'essential place' which exists in its real authenticity waiting to be discovered by the researcher. There is, then, no essential south east.

Even this minimal statement of position already stands in contrast to the arguments of some. Thus, Duncan (1989) has put a case in relation to localities that such places can only claim 'a real existence' if and when there can be seen to be generated within them genuinely emergent effects. That is to say that a place can only claim the status of being a locality if the juxtaposition and interaction of the social relations which construct it clearly result in new, locality-specific, social processes. Of course, such a question may be an important one – and indeed it is one which is both implicit and explicit in all of the chapters in this book. Part of our argument is that what happened in 'the south east' in the 1980s did indeed produce all kinds of socially significant effects and – the point – new social processes, around the functioning of the economy, around class, around gender, around ethnicity.

However, we would not wish, in principle, to restrict admission to the status of 'locality/region' etc. to those places where such effects are generated nor, necessarily, to define their geography by reference to the distribution of such effects. Moreover, and perhaps more immediately to the point, we still needed some notion of 'the south east' as a starting-point from which to ask questions about the possible production of emergent effects.

So where was our region? What was clear from our initial theoretical stance was that the south east did not 'already exist' – we were going to have to construct it as a part of our research. For, as we have argued, 'regions' or more generally 'places' or localities can only be defined for specific purposes, as a result of the posing of specific questions.

Our purpose was to explore 'something called the south east of the UK in a particular decade' (i.e., a particular segment within the wider relations of space–time) as a locus of a particular type of growth. The first move was therefore to examine, in a broad preliminary way, the geography of the – characteristic and particular – mechanisms of growth which we had identified.

Let us begin from a few straightforward national distributions which relate to each of the mechanisms we identified.

Take, for instance, the mechanism of *consumption and credit*. Maps 2.2 and 2.3 show the kind of index which is frequently plotted as part of regional definition. They show the geography of average full-time weekly earnings at the end of the decade. The south east emerges clearly. Yet so, equally clearly, do some other characteristics. The difference between female and male rates is startling (and this is for full-time work only), the highest band for women falling within the lowest for men. It is just one indicator of a host of inequalities. Indeed, in income terms, the south east was the region of the greatest inequality in the country (Forrest and Gordon, 1993). It is not clear, therefore, what averages mean. Moreover, what we are trying to get at is income as part of a growth dynamic – as a propulsive factor. It was the spiralling of top incomes which, we would argue, were more important in generating the circle of growth in the south east in the 1980s, and also which were crucial in determining its nature.

Maps 2.4, 2.5 and 2.6, therefore, show the distribution of high incomes. And again the south east of the country stands out. But here another consideration arises. For it is a different south and east for each of the three income levels (over £30,000 p.a,. £40,000 and £50,000). Map 2.7, unfortunately only at Standard Region level, shows changes in wages between 1979 and 1986. Again, the distribution is slightly different. And it is so again in Map 2.8, which shows regional changes in house prices between 1979 and 1986. These maps all relate to our first-cited mechanism of growth. They allow us immediately to make two important points. First, that different aspects, even of the same mechanism of growth, will display different geographies of the concentration of growth even though for all of them the broad description might be 'the south and east of the country'. Second, that significant aspects of this growth mechanism were clearly present well beyond the borders of the south east Standard Region and even beyond what is more broadly thought of as the south east.

Both of these points emerge even more strongly when distributions are plotted for some of the main indices relating to the other growth mechanisms. Thus, for the finance mechanism, the map (Map 2.9) of 'the Greater City' in terms of its employment and consumption/income effects again shows the same very broad geography but with differences in both wider reach and local intensity. The attempt to isolate the propulsive elements of high technology, by combining growth in high-tech employment with the distribution of qualified scientists and engineers shows, in detail, a different geography again (Map 2.10). So does the distribution of increases in house prices. Finally, some of the operations of national government are illustrated in Maps 2.11, 2.12, 2.13 and 2.14. In each case, the south east emerges as a net gainer in inter-regional distributions over the late 1980s but the geography in each case is different. The geography of other aspects of state intervention over the period further underlines the point. The broad shape of economic policy, favouring (private) services and small firms rather than manufacturing sectors tilted government favour away from the north and towards the south (Martin, 1988), and the privatization of previously nationalized industries led to a major loss of jobs in northern regions (Hudson, 1988). But changes in the defence industries had a geographically more ambiguous effect, leading to losses of employment in some parts of the south east (Lovering, 1991). Changes in the tax regime produced a geography of gain which followed the distributions of high earners and high house prices (Mohan, 1989).

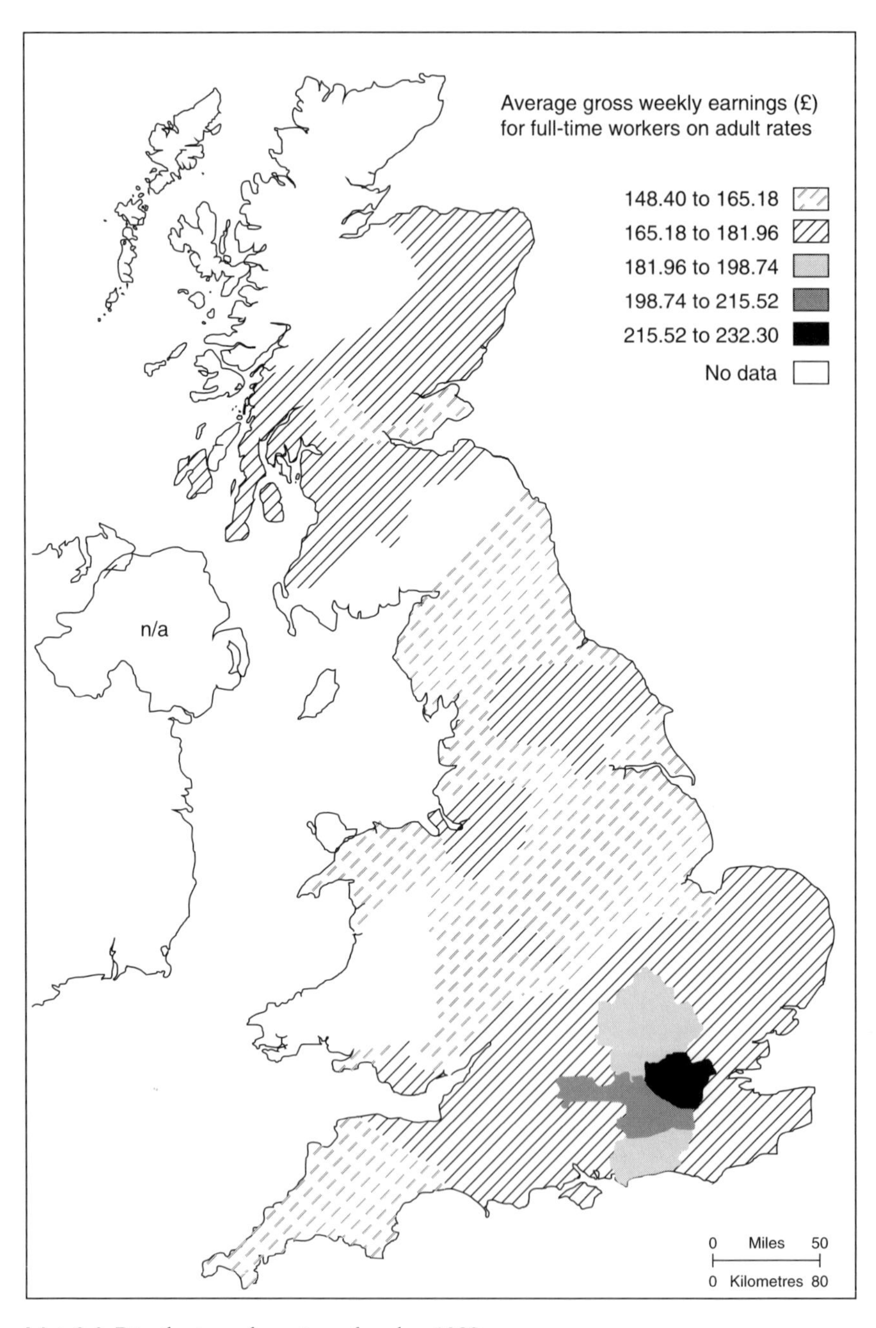

Map 2.2 Distribution of earnings: females, 1989
Source: *New Earnings Survey*, 1990

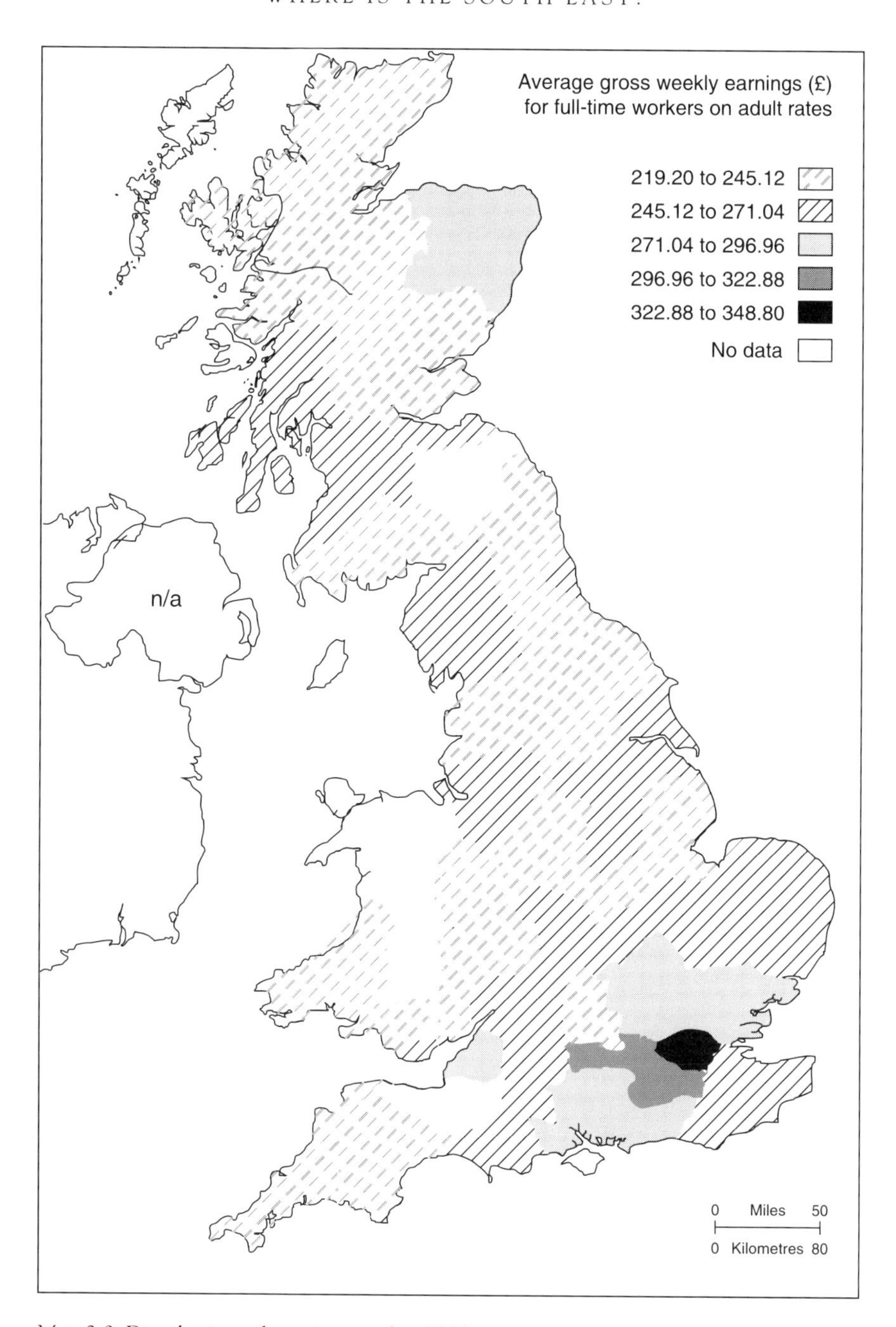

Map 2.3 Distribution of earnings: males, 1989
Source: *New Earnings Survey*, 1990

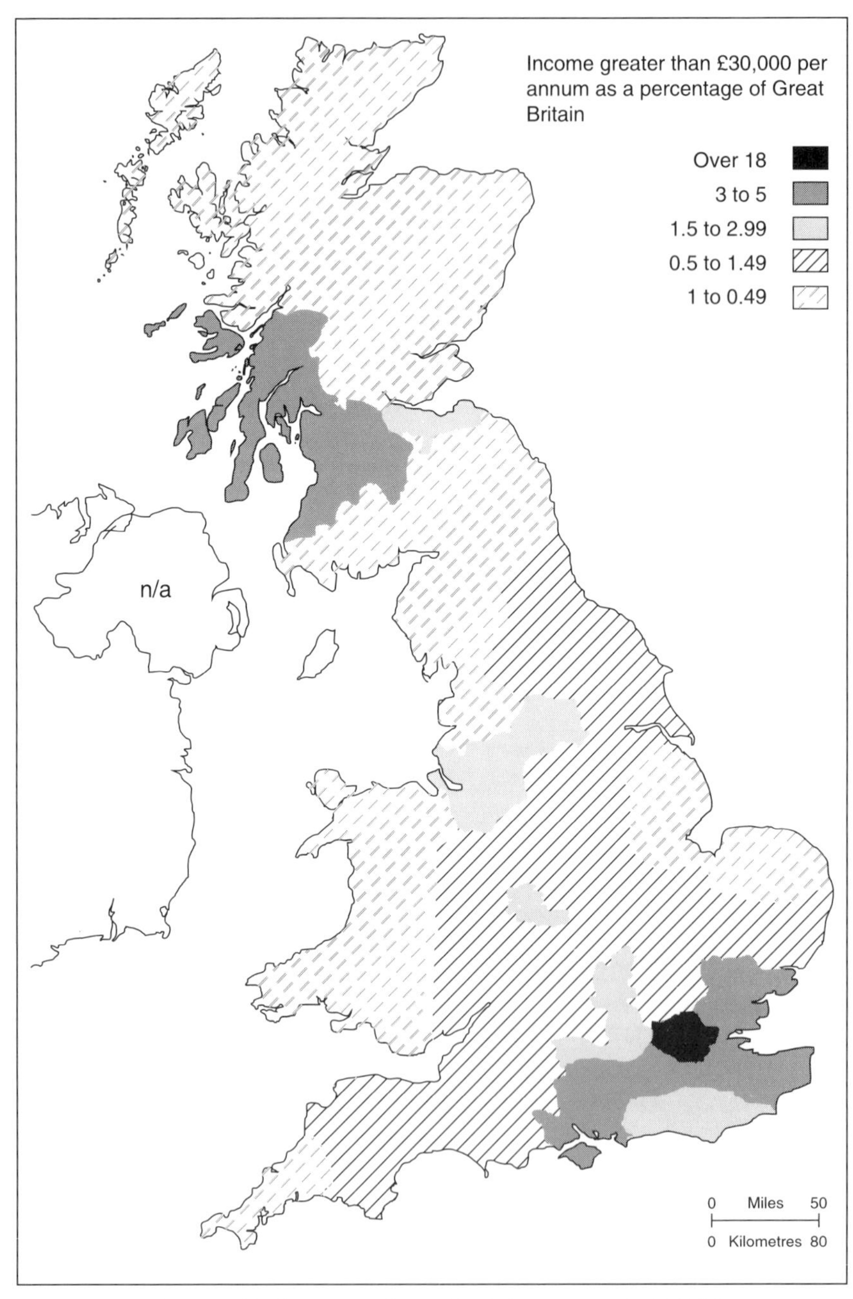

Map 2.4 Distribution of high incomes (over £30,000 per annum), 1989/90
Source: Inland Revenue Special tabs in Hamnett (1993)

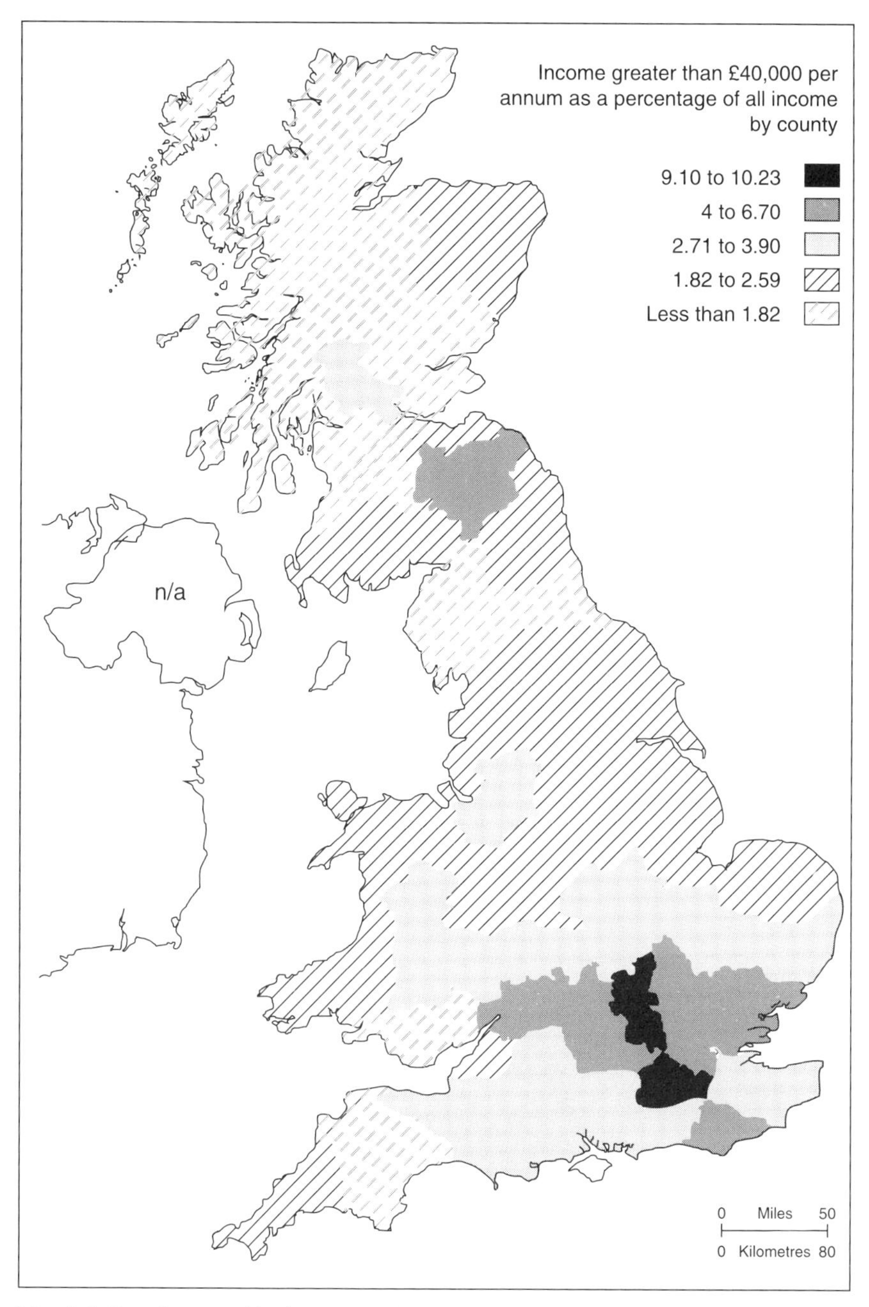

Map 2.5 Distribution of high incomes (over £40,000 per annum), 1989/90
Source: Inland Revenue special tabs in Hamnett (1993)

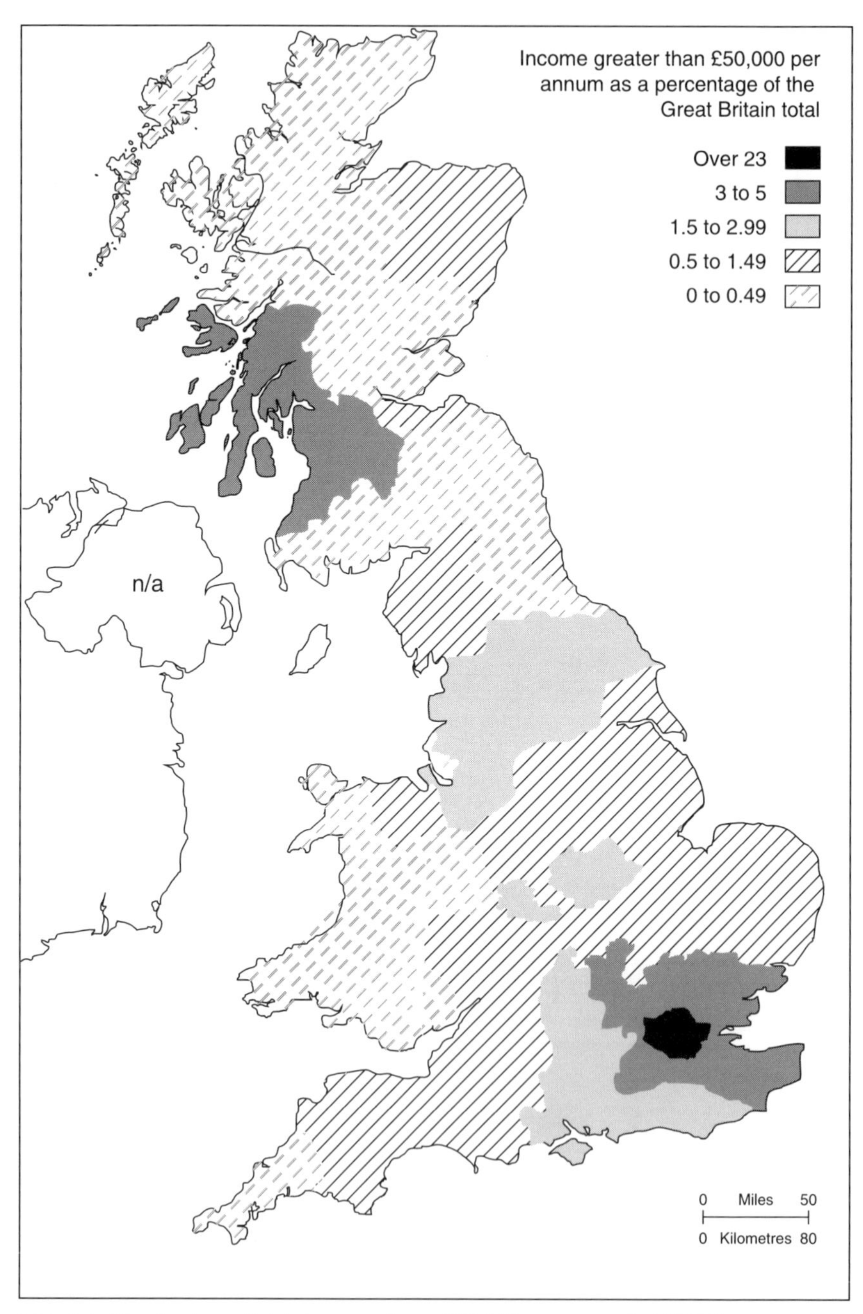

Map 2.6 Distribution of high incomes (over £50,000 per annum), 1989/90
Source: Inland Revenue special tabs in Hamnett (1993)

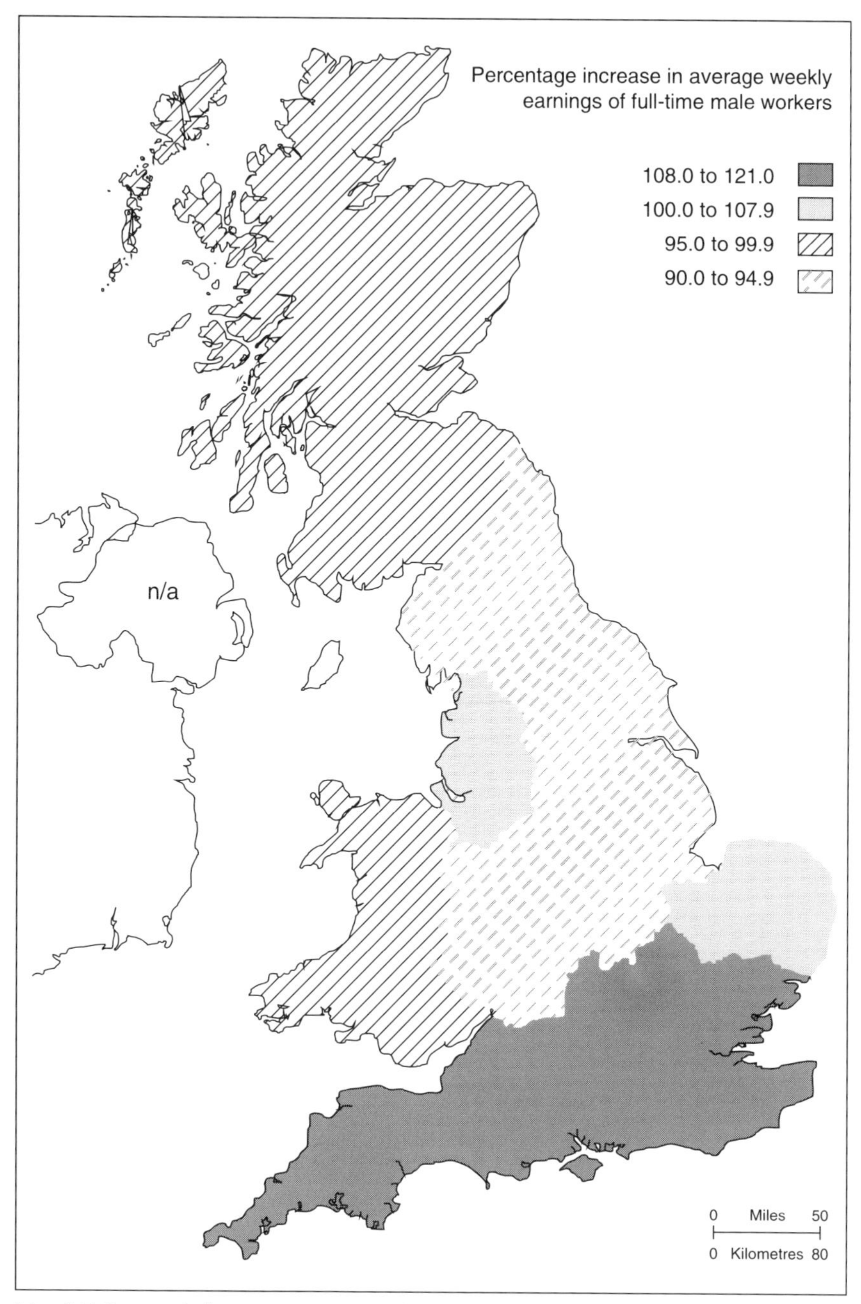

Map 2.7 Regional change in earnings, 1979–86
Source: New Earnings Survey, 1980, 1987 in Martin (1989)

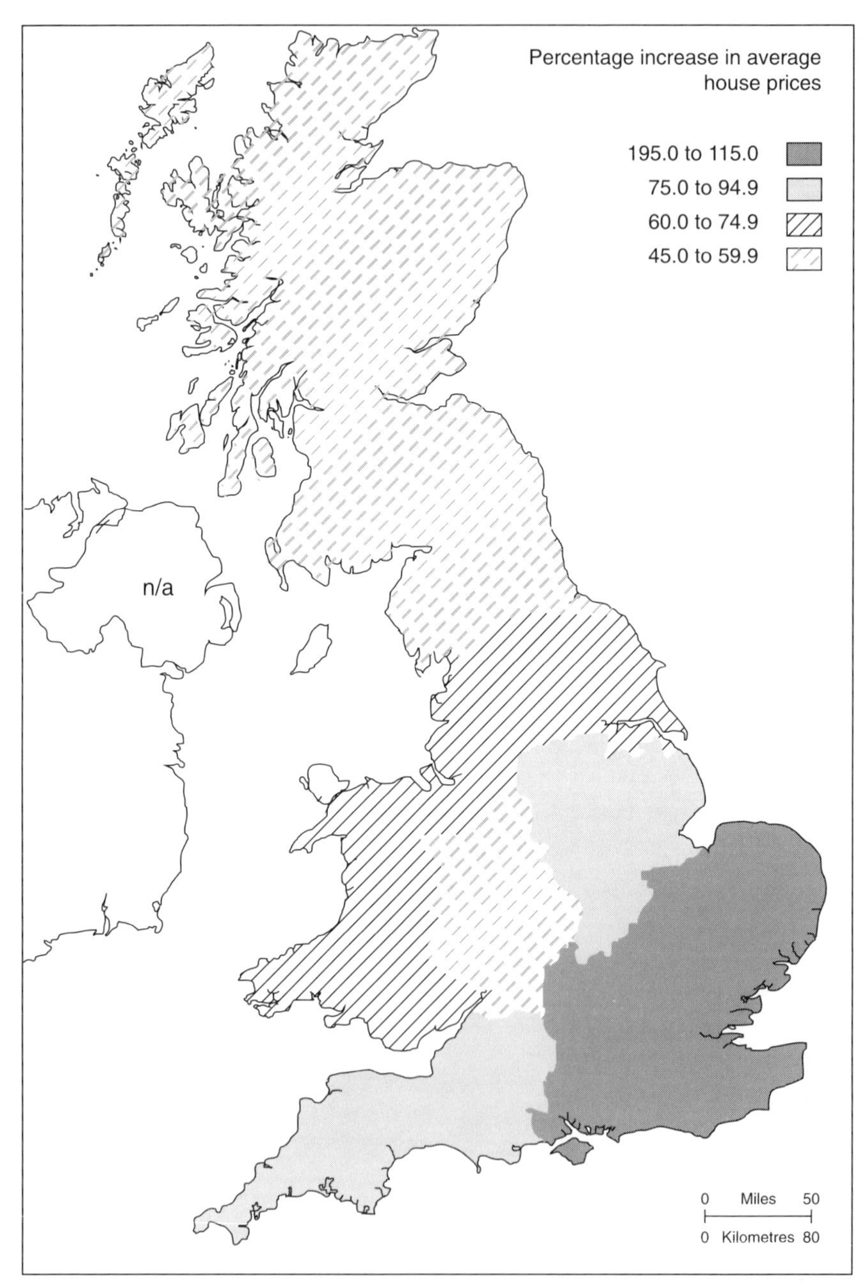

Map 2.8 Regional change in house prices, 1979–86
Source: Department of the Environment, 1988 in Martin (1989)

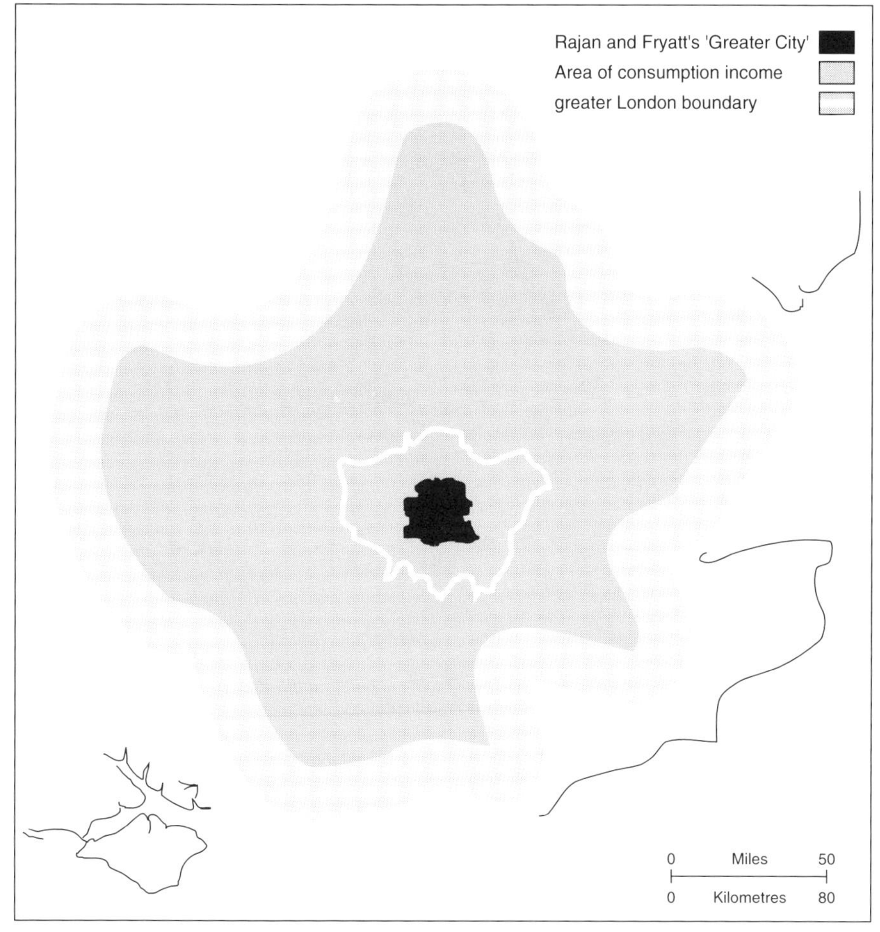

Map 2.9 Finance: employment, consumption and income reach
Source: based on Rajan and Fryatt (1988) and Thrift and Leyshon (1992)

The ROSE even benefitted relatively from the restructuring of the National Health Service, particularly at the expense of London (Mohan, 1995b).

The definition of the region of operation of these mechanisms of growth, in other words, will depend on the measure you choose. This is hardly a new point, nor a surprise. The interesting question is what one does about it. We decided to accept the differences, rather than trying to force or argue them into some kind of geographical consistency. Indeed, instead of seeing it as posing some kind of methodological problem, this very 'inconsistency' can be turned to theoretical and conceptual advantage by itself being posed as a significant object of study. For what this difference in distributions reflects is precisely the usual unevenness of growth: uneven development within and between the mechanisms of growth

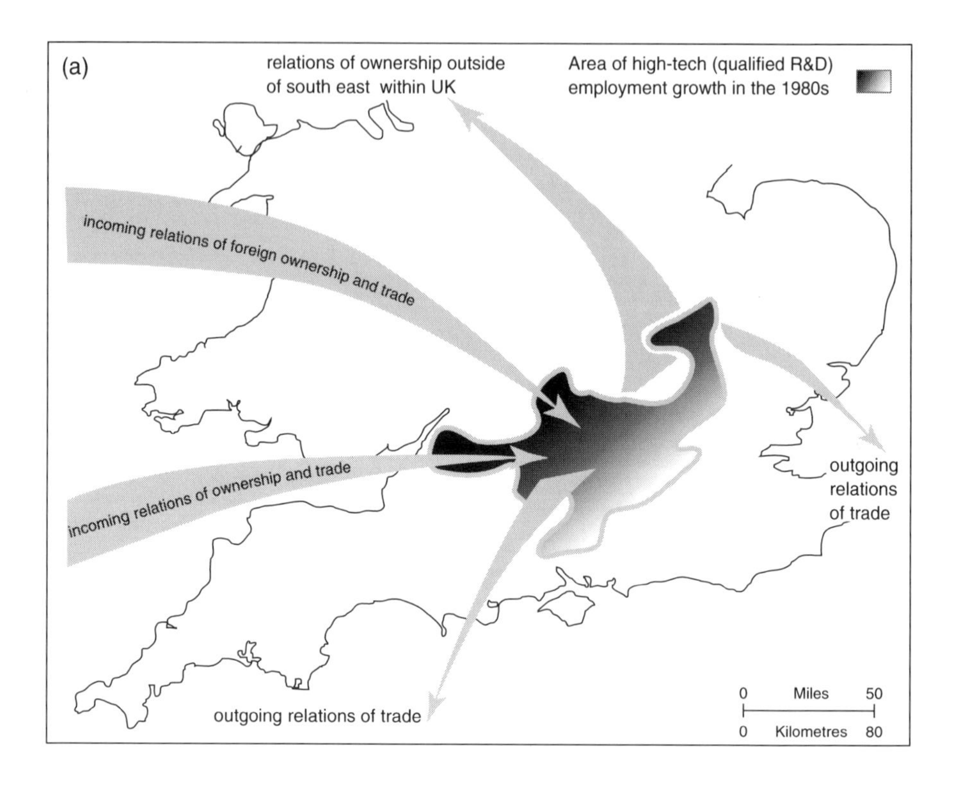

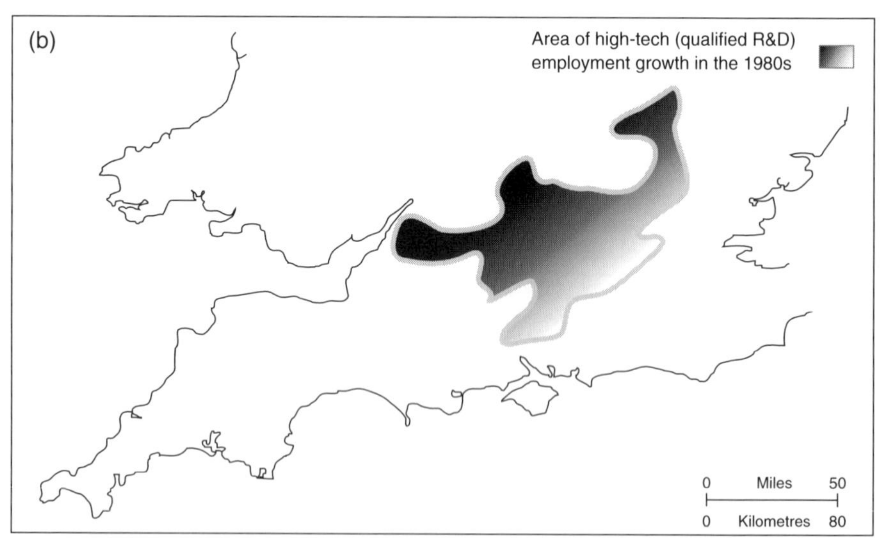

Map 2.10 High-tech employment
(a) High-tech links and distribution of employment
(b) High-tech employment reach

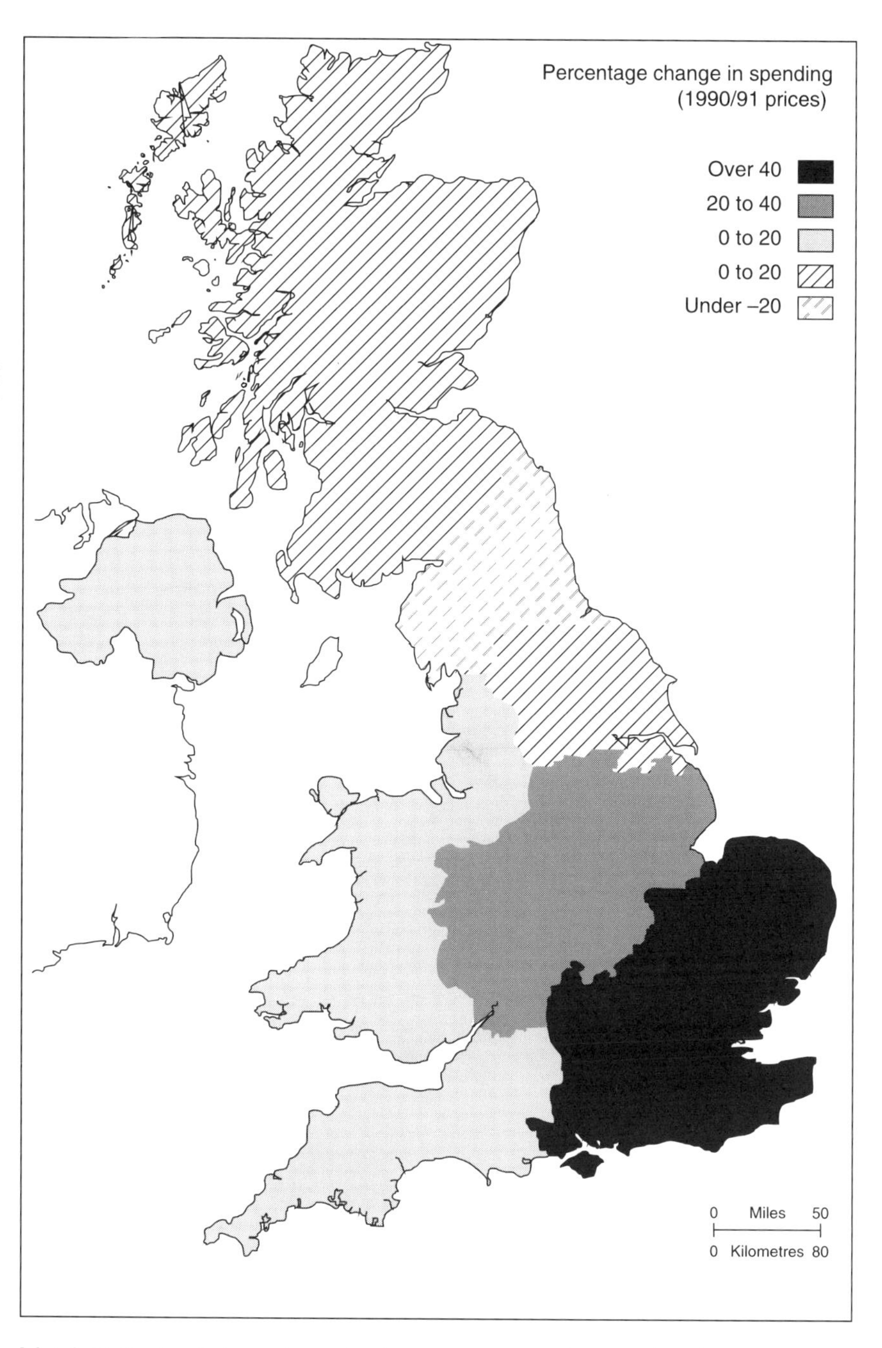

Map 2.11 Regional change in national government spending: roads and transport, 1987–91
Source: HM Treasury in the *Guardian*, 9 February 1993

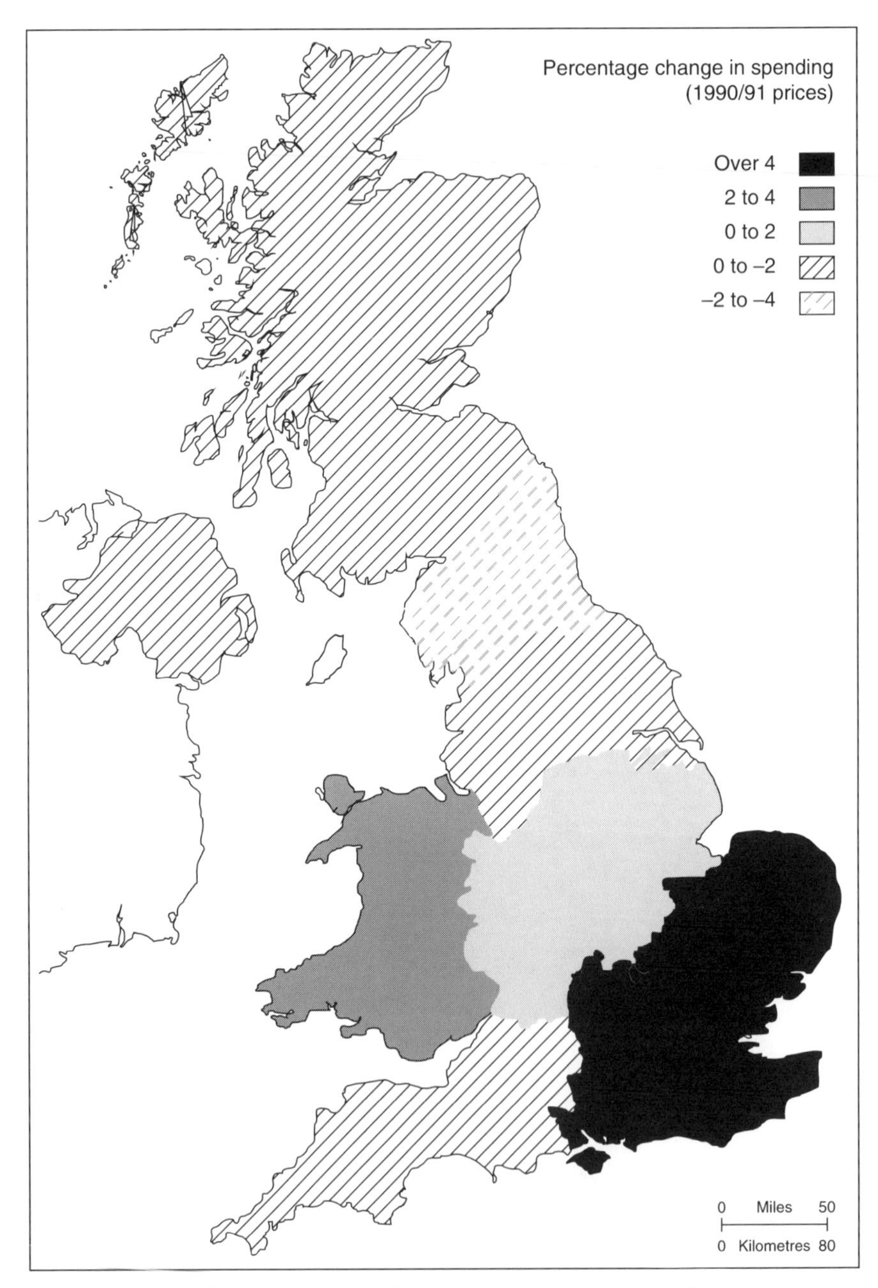

Map 2.12 Regional change in national government spending: regional spending per capita (including defence), 1987–91
Source: HM Treasury and the Ministry of Defence in the *Guardian*, 9 February 1993

Map 2.13 Regional change in national government spending: housing, 1987–91
Source: HM Treasury in the *Guardian*, 9 February 1993

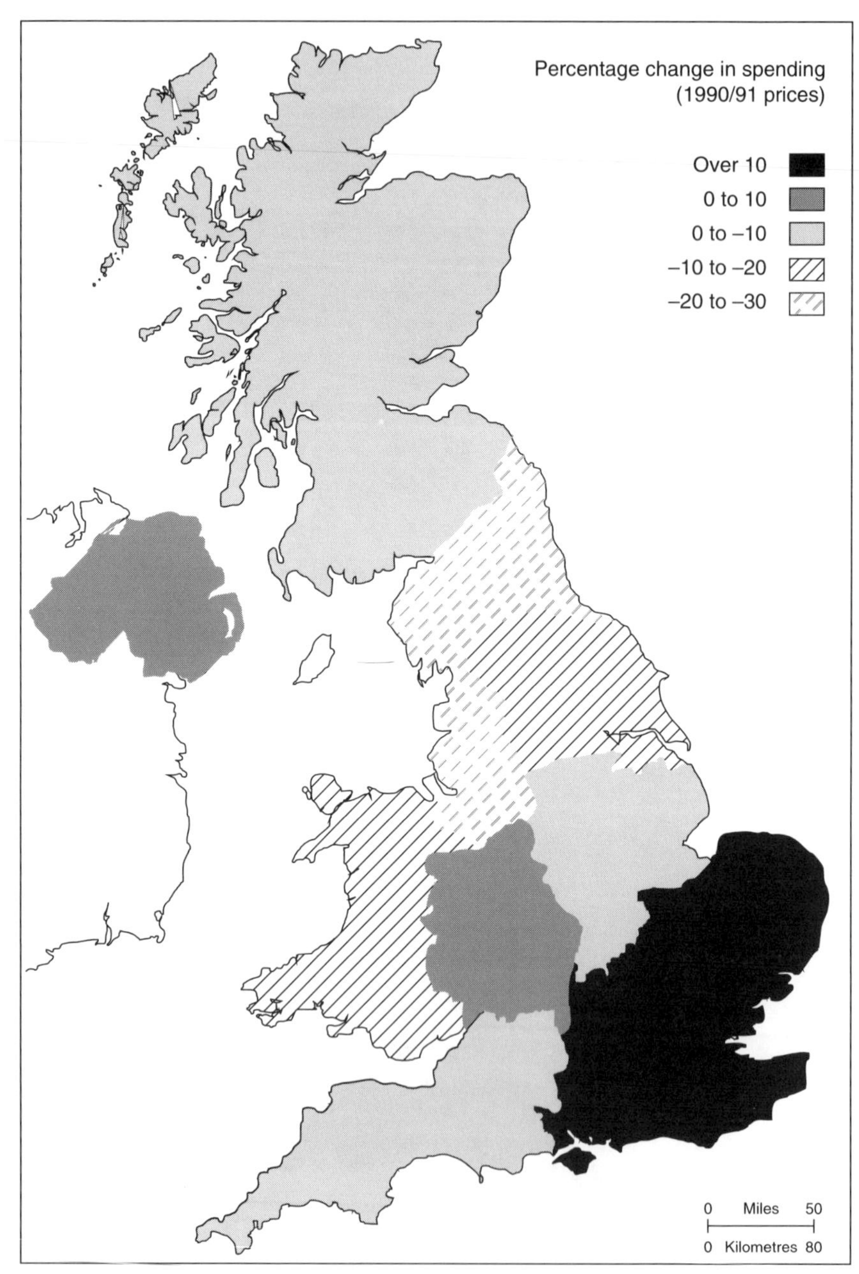

Map 2.14 Regional change in national government spending: trade and industry, 1987–91

Source: HM Treasury in the *Guardian*, 9 February 1993

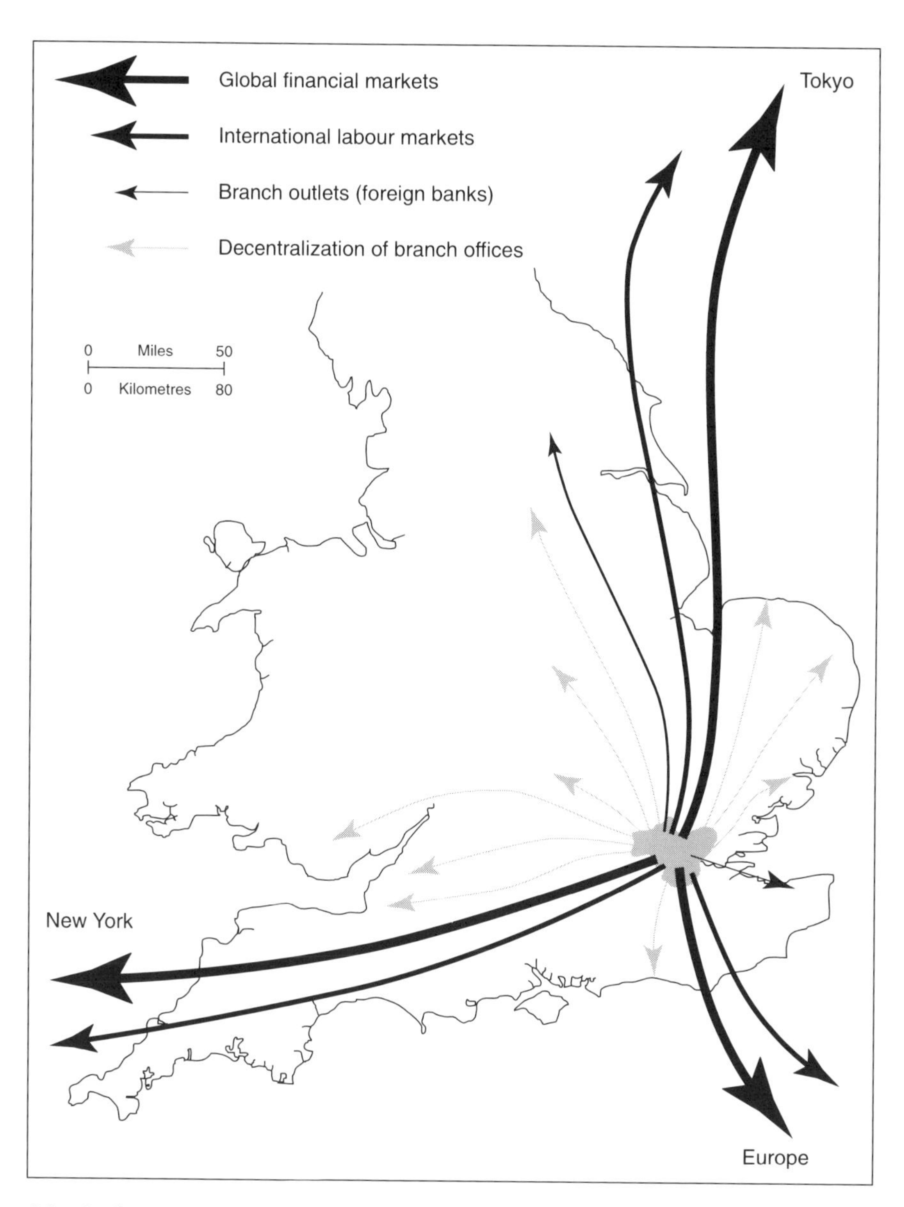

Map 2.15 International financial links

themselves. We decided therefore simply to accept these differences for what they were. Instead of attempting to produce one map of a clearly defined region, we thought more in terms of an overlay of each of these different distributions. This was, moreover, and as will be discussed later, more than an attempt to register a simple additive effect. For it seemed quite likely that the different mechanisms might, where their distributions overlapped, produce at certain scales interactive effects which might further increase the virtuous circle of growth (a form, in other words, of emergent effect). Our definition of the south east at this stage, then, had a number of interesting characteristics. It clearly spread beyond the Standard Region. It was clearly of varying intensity even within the Standard Region. And it was very difficult to say precisely where its edges were.

Moreover, mapping distributions in this way is only one aspect of an approach to the geography of the mechanisms of growth. What such distributional mappings represent is the degree of presence/absence of particular phenomena. Thus, Map 2.9 presents the presence/absence of employment within the 'greater City', and Map 2.10 the degrees of presence/absence of different kinds of high-tech employment. In such mappings, unevenness is represented by differences in level; space is (if only implicitly) understood as a surface; and regions/sub-regions/places are characterized by differences from each other. This is, perhaps, the usual approach to the investigation of regional differentiation.

There is, however, another way of thinking about regional differentiation. For social space is not merely a surface whose undulations represent the varying levels of presence of particular phenomena. Social space, we would argue, can also be understood as the product of social relations. That is to say, it can be conceptualized as the product of the networks, interactions, juxtapositions and articulations of the myriad of connections through which all social phenomena are lived out. On this view, variations over space are conceptualized in terms of differentiated articulations of social relations/processes rather than in terms of continuous but varied surfaces. This is a significant additional perspective, and presents the issue of regional definition in a new light.

To begin with, and at the most general level, it enables a view of uneven development which interprets it not solely in terms of spatial differences in the levels of certain selected criteria (employment, unemployment, per capita income), but also in terms of the inter-regional (including in many cases international) social relations which produce these differences and, at least temporarily, lock them into place.

Second, it means that regions/sub-regions/places are characterized, not by differences from each other, so much as by their place within this overall constellation of forces which constitutes social space and, in particular, by their differentiated and unequal relations to each other.

The example of the growth mechanism of finance will help to illustrate this approach. To begin with, and rather than rushing in to plot every statistic we can find, it is important to consider why we believed this to be a significant element in defining the south east as a region of growth in the 1980s. This has already been mentioned in Chapter 1 but, briefly, banking–insurance–finance was not merely an element in the growth of that period: it was also a *propulsive mechanism*. It was not only an indicator of growth; it was part of what produced it. This was so in a number of ways. In part it was indeed that it was a major generator of employment; and the distribution of this is shown in Map 2.9. This marks an area far wider than Greater London, yet not coincident with the Standard Region, and stretching in particular out to the north and east of the metropolis. However, one reason that employment was itself significant was that levels of remuneration were, for a proportion of employees, very high. This, then, was employment which would feed also into significant increases in demand (and thus link up with the credit/consumption mechanism, and perhaps in turn with that of house prices). We therefore also mapped the consumption/income reach of this sector, also shown on Map 2.9. Here, the spread is geographically wider, since it takes in where people live and primarily spend their incomes; it spreads more to the south and west of the metropolis.

However, these are still distributions – space conceived as a surface of varying levels. But there is more to 'finance as a growth mechanism', as an element in the production of uneven development, than that. It is also a centre of control and a pole of attraction. The City of London is the most important location in the UK for incoming foreign banks. And it is the crucial geographical hinge-point in the whole of the UK for the financial links between the national economy and the world economy. Some of these links are shown on Map 2.15. A geography of input–output relations would reveal another propulsive side to the finance sector and again demonstrate its links to other mechanisms of growth, in particular its close connections with the electronics side of high technology. From the dominant office locations (that is, yet again a different distribution from those of employment and consumption) reach out relations of control, over branch plants and over investments, both in the rest of the UK and internationally. This, then, is a geography of social relations. In this way of viewing things, space is the product of social interaction. This, in turn, means that places/regions are characterized by the nature of the social relations which link them together. And social relations, necessarily, are full of power. So uneven development is not only about unequal levels of distribution; it is about the geography of power. It is this kind of notion of space – and of places within a wider space – which we were attempting to capture as we explored the question: 'Where is the south east?' In this way of viewing things, space is the product of social interaction, uneven development is about the geography of power, and places/regions are characterized by the nature of the social relations which link them together.

If, then, our aim had been to define the south east as a region of banking, insurance and finance, the task would already have been complex. What we aimed to do, however, was to conceptualize the south east as a growth region of the 1980s – that is, by thinking in terms of this kind of complexity for each of the mechanisms of growth individually and for all of them together.

Thus, in relation to high technology the lines of ownership and control run out from the south and east of the country both to other regions and internationally; but they also come in, from the USA, continental Europe and Japan. Input–output flows would indicate a high degree of internationality as well as strong links to the finance sector. The actions of the state have also influenced the geography, not only of 'distributions', but also of power relations. Privatization, for instance, not only resulted in a geography of the nominal control of new share ownership heavily skewed towards the south east (Hamnett, 1992) but also enhanced the expertise and international influence of the City (the finance sector) as the policies came to be adopted in other countries. Likewise, the privatization of previously public services, such as healthcare and residential homes, has resulted in a geography of provision concentrated in, and headquartered in, the south and the east of the country (Mohan, 1988a, 1988b).

It is perhaps useful to pause here and to consider our approach in relation to that of others. In the very last year of the 1980s another study was published of this region, by Breheny and Congdon (1989). It was entitled *Growth and Change in a Core Region: The Case of South East England.* It is an edited collection containing a wide range of papers on 'economy and employment', 'housing and population' and 'strategic planning'. Quite a number of its concerns and not a few of its conclusions chime with our own. Yet in other ways the study is very different. In particular, this is the case with the conceptualization of the region and thus with the range of questions which can be asked. Breheny and Congdon set out the framework in their introduction. The emphasis there is on the south east as a *core* region. Thus, they write (and we would agree very strongly with them in this) that:

> These issues should not be of interest only to UK readers. The south east is regarded as an example of a 'core' region, occupying a dominant position, economically, politically, and culturally, within a capitalist economy which exhibits distinct and enduring uneven development. Other such 'core' regions exist in other parts of the world.
>
> (Breheny and Congdon, 1989: 1)

Yet it is difficult to pin down in what this characteristic of being 'a core region' consists. It is hypothesized that there might have been a change in its nature, or that such a change might be beginning, as a result of an increase in flexibility within the economy and the potential tendency to increased centralization. Beyond this, the focus is mainly on distributions. Thus, 'the growth of the service sector, particularly the producer service sector' is seen as being of especial significance. However, the analysis of this significance is almost entirely conducted in terms of employment growth (ibid.: 3). The establishment of 'coreness' is primarily carried out through mapping differences in indices. On the one hand, this does not elaborate on why particular measures are significant – in what sense they are propulsive in the regional economy. On the other hand the region is merely compared to others. Yet if one region is core, it can only establish its coreness in relation to others which are not-core, and it is real social relations, not merely contrasts in distributions, through which these dynamics of uneven development are played out and held together. Moreover, finally, in their introduction Breheny and Congdon take the boundaries of the south east as given: it is the Standard Region. This can lead to problems with the notion of the south east region as 'core':

> over time the areas receiving the highest growth rates have been further and further out, to the point where now the focus of growth is on those counties immediately beyond the regional boundary.
>
> (ibid.: 2)

and

> Outer areas of the region . . . have been favoured, along with Bristol, which is beyond the south east region but is very accessible.
>
> (ibid.: 3)

The cause of this kind of methodological difficulty, we suggest here, lies in defining separately the region and the characteristics/processes which are said to distinguish it. The solution, or so we argue, is to define them together. If a core region is to be the focus of study then it should be conceptualized in terms of those processes which are argued to make it core.

There is, however, a further issue here. However one defines them, using whatever criteria other than fixed boundaries, regions change. Geography changes with history. Our aim was not only to define the south east in terms of particular mechanisms of growth, but also to do so at a particular period. Moreover, as has already been hinted, the neo-liberal strategy for the economy in the 1980s produced a very particular geography. It was quite unlike, for instance, the geography produced two decades earlier by Wilson's strategy of modernization and social democracy. In the 1960s, although 'the south east' was, as in the 1980s, undoubtedly the richest region, it was by no means so far ahead of the rest of the country as it was two decades later. Within the south east the patterns were different too: there was less marked inequality between social groups and between places. And the boundary between the 'north' and the 'south' was different. In particular, the Standard Region of the west midlands was in the south – it was one of the regions of growth. The different mechanisms of growth which were emphasized in the 1960s – manufacturing rather than the City, public sector services rather than private, labour market regulation rather than its opposite, high technology promoted through merger rather than the market – produced a different geography. The symbolic meaning of 'the south east' was different, too. The political bases of the government of Harold Wilson lay in the north and west – in Scotland, Wales, the north of England. The valorization of manufacturing gave these places a positive image (even while their coalfields were being run down) quite unlike the dominant image of them two decades later.

The region we were trying to define, therefore, was a particular space *at a particular time*. The 1980s' decade of neo-liberal politics *created a new south east*, and it is this south east which we set out to explore.

Further theoretical and conceptual considerations

This way of approaching regional definition raises a number of important issues. Moreover, these are not 'only' theoretical issues, in the sense of being our own private obsessions of conceptualization. They matter to how the research is done, to the range of questions which can be asked, and to the coherence of the conclusions.

First, conceptualizing regions in this way means that they remain essentially unbounded. We never drew boundaries between our south east and other regions of the country. In this sense our answer to the practical questions of geographers about drawing lines around regions/localities is that it may not always be necessary. Certainly, boundary drawing would do violence to this kind of conceptualization. This is true for a host of reasons. There is the necessary fuzziness which results from overlaying the different geographical distributions of the selected identifying criteria. There is the fact that difference is here constituted more through interconnection than through opposition. And there is the fact, which will be explored below, of 'intra-regional' differentiation and even discontinuity.

The south east as we think of it is not bounded off from but linked to, and in part constituted in its character through its linkage to, other regions. This is clearest in relation to the interactions between activities in the south east and the rest of the UK. It is also clearly apparent in international links, and in the specificity of the nature of those links which distinguish this part of the country from other regions. Many economic activities in the south east form a link between the extra-UK economy and the national economy. There is also a sense in which parts of the economy of the south east form an international enclave within the wider UK space (see Allen, 1992). This approach is thus closely allied to the notion of the uniqueness of place being constructed in part out of sets of interdependencies with elsewhere. It is also, in fact, a little-appreciated corollary of the widely accepted fact that 'you can't explain a place by only looking inside it'.

None of this of course means that there are no lines or boundaries in social space. But – like all the other relations which together form social space – they are social constructions, put there for specific purposes and within particular sets of power relations; they are in principle contested, and they may be used in the course of social contests. They may also, of course, have effects. But such effects must be a matter of case-by-case empirical investigation, for their significance and nature can vary. Some boundaries, such as national ones, may in general have powerful social implications, yet even here there will be variations. Within the EU at present, for instance, some aspects of those implications are being reduced in importance, while the boundary around the EU itself is being constructed, as part of a political process, to have increased significance. The boundaries of local government may or may not be important in the shaping of social space. Thus, for one of the places within the south east which figures in our research – South Buckinghamshire – boundaries and their definition are of signal importance to its very constitution and identity; the people of South Buckinghamshire, as will be seen, defend themselves and their sense of self behind their borders.

The boundary of the Standard Region – the south east – however, and other associated and established ways of defining the social space of the region (such as SERPLAN, the planning region) do not seem to us to produce significant effects. They induce neither regional coherence nor differentiation from areas 'beyond the boundary'. (Perhaps their main effects are on statistics and research – and these effects, the ones we are precisely trying to avoid – can be powerful.) Had they done so, had they been important mechanisms in the construction of social space, we would undoubtedly have been far more reticent in de-prioritizing them in our definition of the region. For they would indeed have had meaning. This lack of effect may be linked to the much lesser general significance of 'regionalism' within the UK, and especially within England and particularly the south of England. In Germany the boundaries of most of the Länder not only reflect long-established economic and cultural differences they also, because of the greater degree of decentralization and regional autonomy in that country, have greater effects. The high degree of centralization within the UK economy, and the political centralization itself reinforced during the decade which we are examining, thus had an influence on our regional definition in practice. This is an issue, then, which runs from regional conceptualization to questions of policy, and it is considered further in Chapter 5. The point, however, is that the construction and definition of boundaries therefore represents one of the many social processes which may – or may not – contribute to the character of an area.

Second, the kind of region which emerges in this approach to conceptualization may not be spatially continuous. It may have 'holes' in it – that is to say areas 'within' the region which are not characterized by the mechanisms/features which are part of the criteria for regional definition. Again, this is a situation which arises in other approaches to definition too, but which is usually seen as a problem, recognized in the write-up, but then simply erased by those areas simply being included (as unfortunate non-conforming oddities) within the boundaries of the region.

Thus, in the case of the south east there are areas, clearly within the compass of what would normally be named as that region and surrounded by areas of expansion, but which, equally clearly, were not touched directly by any of the mechanisms of growth which we identified. A number of them – Thanet, Sittingbourne, Sheerness, Harwich, parts of London, along with others – by the early 1990s featured on the map of regional assistance. As in the case of 'inconsistency' in distributions, our approach has been to treat the existence of such areas, not as a methodological irritant to be got around, but as a theoretical/ conceptual issue.

For such areas do indeed raise thorny conceptual problems. In particular, they raise the issue of mechanisms of inclusion and exclusion which may be inherent in the kind of growth which took place in the 1980s in the south east in general. Thus, the Medway Towns seem to have been barely touched by that growth, even though they are, locationally, set within the region. Except for outmigration to London, their links in to the rest of the region are relatively weak. Indeed, politically, and maybe in part in some response to that lack, they have turned to Europe for contacts on which to base a strategy for the future.

There are two possible ways to understand such a situation. In the first case, if the definition of social space in terms of interrelations is adhered to, along with the concept of region as defined by the spatiality of the operation of specific mechanisms, then the Medway Towns, excluded from growth, can not be defined to be part of the region. It is this phenomenon which gives rise to discontinuous regions – regions for which we coined a new metaphor: the doily, for they have holes in them between the connecting links.

This is not the end of the issue, however. For it raises the question of the nature of the growth which took place in this period. Is the fact of holes within the region, (and – in the case of this form of growth – the consequent existence of severe intra-regional spatial inequality) constitutive of growth of this neo-liberal kind? Such questions demand attention both to the conceptualization of growth and to the specifics of each case; and they will be explored in Chapter 3.

Third, apart from actual spatial discontinuities there will also be intra-regional variation in regions conceptualized in this way because of the uneven nature of the overlay of the different criteria (here, growth mechanisms). There will be variation both in the intensity of growth and in the mix of growth mechanisms which are articulated in different parts of the region. And virtuous articulations may produce further, emergent, growth effects. Thus, in contrast to the holes in the doily, other local areas may be 'hot spots', where a particularly strong mix of mechanisms is articulated together.

Thus, within the south east, Cambridge is one of the strongest areas of growth of high technology, and a centre of power and status in terms of perceived technological advancement. We further hypothesized that, because of this, it might show up relatively strongly on the mechanism of consumption/house prices. Further, it has, as will be seen, strong elements, also, of munificence on the part of the national government. Other local places have different mixes/ articulations. The City is, of course, overwhelmingly dominated by banking/ insurance/finance; Fulham and Hammersmith by house prices/consumption, as are Chiltern and South Buckinghamshire. Each place, in other words, is the focus of different mixes in the overall regional space of social relations.

Given this unboundedness, spatial porosity and internal variability it is interesting to consider, fourth, how much this way of conceptualizing regions/places indicates the existence, or conforms with the way of thinking, of structured coherence. 'Structured coherence' has been proposed, particularly by Harvey (1985), as a characteristic, indeed necessary, geographical tendency under capitalism, 'a *structured coherence* to production and consumption within a given space' (ibid.: 146; emphasis as in the original). Thus:

> There are processes at work, therefore, that define *regional spaces* within which production and consumption, supply and demand (for commodities and labour power), production and realisation, class struggle and accumulation, culture and life style, hang together as some kind of structured coherence within a totality of productive forces and social relations.
>
> (ibid.: 146; emphasis as in the original)

To this formulation, the notion of regional class alliances is also subsequently integrated. Although equally necessary countervailing tendencies are indicated, which can lead to their constant undermining, such structured regional coherences are none the less asserted to persist empirically. Moreover, it should be noted, the term 'regional' does not limit the relevance of this concept to any particular geographical scale. Harvey himself works the concept at a whole range of scales including national, subnational, regional and smaller community.

There are a number of points of common emphasis between the notion of structured coherence and the approach to regional conceptualization adopted in this book. Both are concerned to think of space in terms of social relations, both indicate notions of region/place as essentially porous and unstable, subject necessarily over time to spatial reworking. Moreover, our own way in to the study of the south east indeed had as its starting-point a kind of geographical coherence in the sense that the growth mechanisms we identified, both individually and in their interlinkage, have a coherent regional geography simply in the sense of not being just scattered all over the country. The south east was clearly a locus of power. As was pointed out at the beginning of this chapter, this regional geography was indeed a significant element in the specificity of 'free-market' growth.

There are other ways, however, in which our approach to conceptualization differs from structured coherence. Most particularly, the 'open doily' approach does not imply the necessary inclusion within the definition of a region of everything which lies within a spatially continuous area. Thus, first, all across the region which we have defined as the south east there are economic activities and most certainly aspects of culture, lifestyle and class, which it is difficult to see as coherently integrated. At one level this reservation implies an in-principle lesser commitment to determination by the economy than is suggested by the usual formulation of structured coherence. At another level, however, such degrees of coherence between aspects of society must be – we would argue – a matter of empirical investigation in individual cases. Such investigation, along with exploration of the nature and strength of class and other spatially based alliances, will form one element of what follows. But the second significant difference between our concept of region and that of structured coherence is of course that our regions may have, indeed may be likely to have, holes in them. The regional coherence we identify, in other words, is that constructed by certain economic relations; it is not a coherence which links together all which lies within an absolute space marked out by coordinates of latitude and longitude. This particular 'coherence' moreover may overlap with, intersect with, even exist in a contradictory relation with, other geographies of other social relations. What we are working towards, in other words, is an ability to define regions/places for certain purposes (in order to be able to ask certain questions) while at the same time holding on to the notion that this is one way of viewing, one perspective on, a fuller spatiality which may be full of incoherences and paradoxes.

Fifth, none the less, in spite of this emphasis on the possibility of incoherence and paradox – indeed maybe simply as its other side – it is also important to analyse the nature and degree of integration of this hypothesized region the south east. Again, the approach of thinking of space in terms of social relations is helpful, for what is at issue here is not so much the degree of similarity between constituent places but the nature and degree of their interconnection. A particular question is the relation between the outer south east and London. In many ways, as later chapters will explore, the two exist in a material and symbolic connection-through-counterposition. This is especially true socially. London, 'the urban', and all that that is thought to entail, is part of the reason for decentralization and one of the bases for the construction of the character of the outer region (Cloke, Phillips and Thrift, 1995). Imaginations of race and constructions of ethnicity are, as Chapter 4 argues, particularly important here. But what of the functional connections of the economy, culture and social life? Is the outer south east essentially a vast suburbia with its critical arteries leading to and feeding off London, or does it have, either as an integrated place or as a set of more separately functioning units, a significant degree of autonomy from the capital? This last scenario is the one painted by Garreau (1991) in his argument that today such areas are emerging as 'edge cities'. It is a hypothesis we explore in Chapters 3 and 5. Given the geography of recent growth (see the discussion of Breheny and Congdon, above) these are not issues which can be explored within the confines of the pre-given boundaries of Standard Regions.

Sixth, although the programme of research was concerned with the south east, in fact individual projects within it each focused on one or more smaller-scale localities (Map 2.16) usually at the level of towns (Cambridge, the Medway Towns) or roughly the size of boroughs (South Buckinghamshire, Hammersmith and Fulham). This level of focus corresponded, on the one hand, with a desire to capture the variability within the south east, and, on the other hand, with the fact that the processes we wished to study in depth were hypothesized primarily

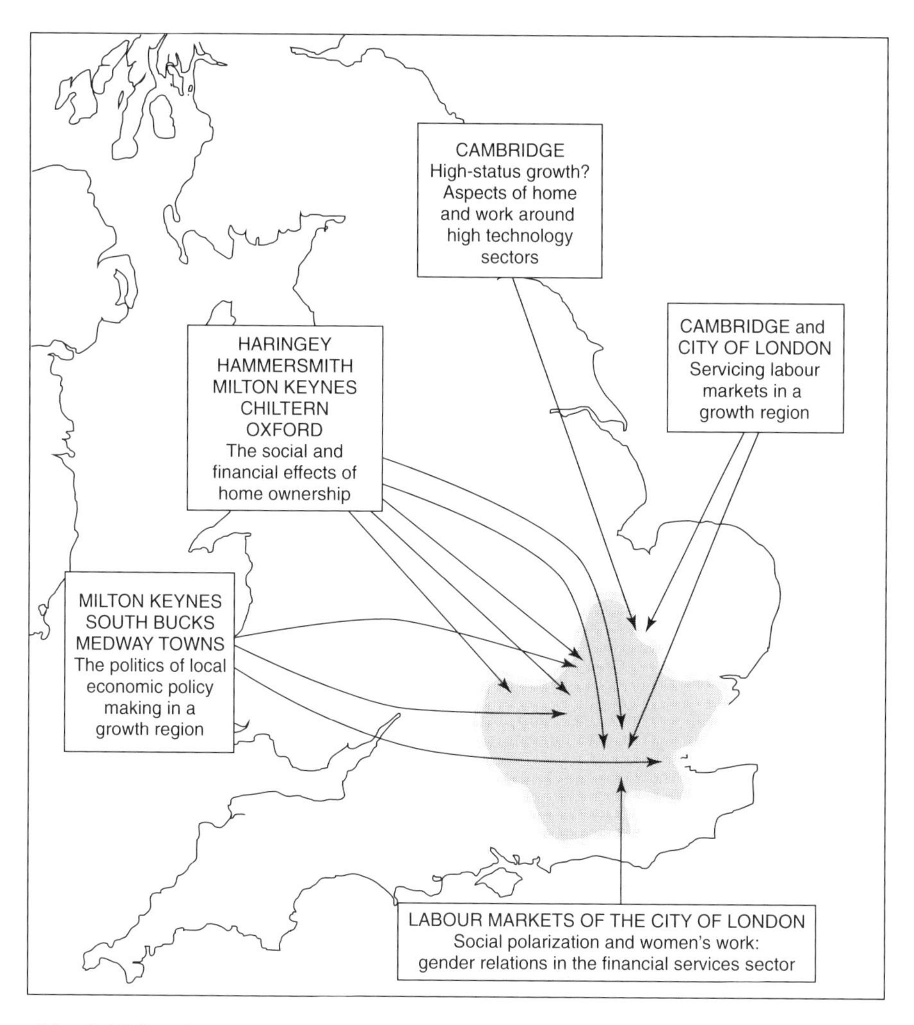

Map 2.16 Local project areas

to operate at or below this spatial scale. Now, these local areas are conceptualized in principle, and empirically defined in practice, according to exactly the same tenets as applied to the regional level. Like the region, therefore, they are open and unbounded. Their individual specificity and their differences from each other are seen as arising in part from their differential constitution as contrasting mixes of growth dynamics (see above) and in part through the different ways in which they are interconnected – to each other, to the wider region, and to the world beyond. One immediate advantage of this approach in research terms is that each local area forms an 'arena' for the investigation of potential interactions between the particular mechanisms through which it is defined.

This stress on the definition of local areas on the basis of the intersection of particular bundles of growth mechanisms to some extent differentiates our approach from that adopted by Cooke (1989) for the definition of localities. Moreover in the series of locality studies to which that definition was linked, the questions at issue, and therefore the prime criteria of definition, differed from our own. Thus, the localities projects, beginning from economic restructuring and searching out its effects in the labour market, in daily life and in political proactivity, took each of these as prime criteria of definition. In those projects, localities were circumscribed, and circumscribed in part by labour market and in part by local authority boundaries (local authorities being seen as the prime agents of local proactivity – or the main ones to be investigated). We do not follow this route, in part because our questions and therefore our criteria are different. Also we did not want, nor did we need, to do strict statistical comparisons between sets of areas. In principle, we could afford to keep our local areas, as our region, open and unbounded. Yet at some points the questions asked, as in the localities projects, required a definition of agents which in turn implied a recognition of boundaries. This was the case in particular in the study of local authority activity in the context of the highly varied pattern of growth and absence of growth within the region as a whole.

But there is a further implication entailed in this way of conceiving of local areas. Clearly, in one sense, we are dealing here with processes operating at different geographical scales – we have already in this chapter referred to the international, the national, the regional (south east) and the local area. And clearly there is a social reality to each of these levels. And yet, we would argue, seeing things simply in this way is highly restrictive. As has already been argued, different kinds of social process have very different geographies and they do not all fit neatly into the same set of nested hierarchies. It is from this that derives part of our reservation about the concept of structured coherence. Distinct social processes have more geographical autonomy than this formulation implies. But, further, even within any particular set of social relations the different spatial scales are constitutive of each other. This means more than that processes do not work in a 'top-down' way, that – say – local phenomena may influence wider processes, and so forth. It is a statement about the construction of the local, the global, the national in a mutual way, rather than seeing the 'global', for instance, as the result of adding together a lot of little 'locals'. However, if this is so then the global cannot be treated as merely a scene-setting backcloth. Thus, one frequently adopted way of doing regional or local studies is first to sketch in, say, the international situation and its influences, then the national, then the 'regional' . . . proceeding on downward through spatial scales until the object of study is arrived at. It is what might be called the 'Russian doll' approach to area studies. Against this, we argue that the different levels must be conceptualized together from the start and the empirical work must, where necessary, move back and forth.

Finally, a methodological point. Our approach to conceptualization does of course pose statistical difficulties. On the one hand, if there are no essential regions on what basis are standardizable statistics to be collected? One possibility is for them to be collected at an extremely small scale from which they can subsequently be aggregated. But there are all kinds of practical problems with such a procedure. Thus, for general purpose interregional comparisons, the Standard Regions may be as good as anything else. The point is that we as researchers should not necessarily use them; we should if necessary or appropriate construct our own. Thus, as one example, our investigations of the economy of Cambridge used statistics for the City, for the County, for the Training and Enterprise Councils, for special surveys, and mixed and matched and juggled them as occasion required.

In spite of these difficulties, however, and as should become evident, this approach to regional definition enabled us to cope with change over time, to explore the internal structure of the region and variations within it and to analyse its relations with other regions which, as will be seen, were to prove an element in the ultimate fragility of the growth of the 1980s.

Part II

REGIONS AND IDENTITIES

3

IDENTITY OF PLACES

The growth which shaped the 1980s in significant ways produced a new south east, a new space–time. What was this place like 'inside'? Far from the image of a bounded area on a map (the way in which we customarily imagine regions), the approach adopted here reveals a region which is by no means a 'whole', with all the characteristics of coherence which that term implies; nor is this region a bounded entity. Thinking 'a region' in terms of social relations stretched out reveals, not an 'area', but a complex and unbounded lattice of articulations with internal relations of power and inequality and punctured by structured exclusions. At one level this is a general argument for how we should conceptualize any place, any region. But at a more substantive level this analysis also reveals aspects of the detailed production of a particularly neo-liberal form of uneven development, even within a region of its claimed success.

A (London) region?

One of the more enduring ways in which the south east has been structured into a single region in the popular (and academic) imagination has been through its focus on London. This view of it – though one which has frequently been problematic for planners – has provided it with a sense of wholeness.

There are undoubtedly ways in which this view accords with important aspects of the geography of social relations. Simple things – like the overwhelmingly centripetal pattern of commuting, the dominance of London as a cultural focus, the clear processes of decentralization – provide immediate substance to this image. Even the equally clear evidence of attempts to establish symbolic *difference* from London, and of that being part of the motor for decentralization, can be read as a kind of unity-in-tension in which relations within the region and focusing on London provide the overwhelmingly dominant principle of spatial structure.

And yet, increasingly, the evidence is that this image is inadequate. The south east today is far more than 'the London region'. This refashioning, and increasing complexity, of the space of social relations is by no means only a product of the 1980s, but there is – equally – no doubt that that decade gave the process an additional impetus. What has resulted is a wide stretch of country around the capital parts of which exhibit significant degrees of autonomy from London and within which there is a considerable complexity of internal relations. Is this, then, one of Garreau's (1991) edge cities? Is this the form which the new 'city-states', increasingly talked of as the emerging dominant spatial form of the advanced capitalist world, will take?

One aspect of this growing complexity and decentredness of the region can be detected through the concept of 'hot spots', introduced in Chapter 2. Hot spots are foci of growth, conceptualized as particularly strong points of overlap (and probably also of virtuous articulation) of mechanisms of growth. In Chapter 2, Cambridge was identified as such a node. Here, the growth mechanisms of high technology and government preference intersect and interlock, backed up by reverberative growth through consumption and property. Now, the inclusion of Cambridge in any definition of the south east requires some thought: it is beyond the boundaries of the Standard Region. Its inclusion in the first place, therefore, results from the strategy of defining the region through growth processes. The contrast here with Breheny and Congdon (1989), for instance, has already been noted in Chapter 2. Moreover, the identification of Cambridge as a hot spot, where growth mechanisms significantly overlap, means that its growth is not being considered as necessarily, or primarily, resulting from some kind of spread-effect from the capital. This latter is the usual interpretation. Our view, however, is that the growth has more autonomy than that, that the intersection there of strong dynamics of growth was not simply due to a ripple effect from London finally lapping over into east anglia. Further, while there certainly are relations with London and perhaps especially with the City (venture capital moving one way, high technology the other), Cambridge's most significant positioning in the spatial lattice of growth dynamics is less in terms of its relationship to London and rather more as one node in the set of high-technology locales which includes also Silicon Valley and Tokyo.

There are other ways, too, in which an approach through the geography of social relations reveals a south east which is far more than a stretching-out of relations with London into a 'beyond' of the wider region. We can, for instance, detect strong relations between places within the south east, relations which do not pass through London but do contribute to the formation of new, and the reinforcement of existing, place identities. Within the south east we can see evidence of 'places' 'using' each other to achieve their goals. In terms of development, for example development directed to one area in exchange for lower levels of development in another, it is clear that planned new towns have in a sense traditionally fulfilled this role *vis-à-vis* London. But the process may extend beyond this. With the substantial pressures to find locations for new development of housing and commercial space taking place in the south east during the 1980s, conflict inevitably arose between areas largely perceived as pro-growth and those apparently anti-growth. The last of the planned New Towns, Milton Keynes in Buckinghamshire, came to be viewed as a place to redirect unwanted growth pressures not just from London, but other parts of the south east and beyond. To some extent, other south east New Towns had started to fall out of favour as locations for new investment, being associated with older rounds of manufacturing investment; Milton Keynes promoted its image of a dynamic and welcoming environment, home to the 1980s growth industries: high tech, service sector, foreign multinationals, etc.

This type of relationship, that of interdependence between places, can be identified as exceptionally strong within the same planning authority: the example of Buckinghamshire in this chapter illustrates how the south of the county was able to use Milton Keynes as a means of preserving its dominant identity as a rural haven both from London and from expanding neighbouring districts, ensuring the growth of property prices while also maintaining good public transport links to London (Murdoch and Marsden, 1994). This kind of thing was not confined to Buckinghamshire: all over the region, attempts to preserve a notion of south east 'rurality' were being forged (see ibid.; Marsden *et al.*, 1996). In Berkshire, conflicts arose between Conservative voters in favour of conservation and Conservative interest groups – landowners and employers – seeking further development (Barlow and Savage, 1986). This shows that, despite the nonsense of many local authority boundaries in the south east, local political actors and residents can still use them to maintain their identities.

Therefore, understanding the growth of the south east requires a consideration of what was happening in places outside London and the nature of relations between them and beyond the region. In common with regions elsewhere, particularly north America, new developments – whether 'edge cities' (Garreau, 1991) or 'stealth cities' (Knox, 1993) – are taking place on the outskirts of core cities, and existing places continue to struggle to find elements of distinctiveness which will aid their pursuit of autonomy and prevent their designation as 'non-place places' (Zukin, 1991). In the case of the south east, the identities of such places have become in many ways distinguishable from London's. It seems as though some social groups and places are increasingly trying to 'distance' themselves from London, partly through these places not wanting to be London. Therefore, Garreau's (1991) suggestion that 'edge cities' – new self-functioning settlements – may exist around London (in common with Los Angeles) and provide a form of escape from the more threatening aspects of metropolitan life, provides one way of starting to understand the nature of places in the south east. However, he regards their identity as constitutive of more than an expression of tension with the urban core. His characterization of London (meaning the south east) as 'the most America-like urban centre in Europe' (ibid.: 108) is therefore potentially quite convincing.

There is no disputing the importance of London in creating the south east's key position in the discourse of dominance in the regional system of representations of which we spoke in Chapter 1. But it would be a mistake to assume that London and the south east are interchangeable terms for the same place. One aspect of our approach is to examine the nature of the social relations between places which have resulted in spatial variation. The notion that all places in the region are locked into relations with London, and thus take the prime axis of their identities from that relationship with London, is contested here. Place-identities within the south east of England are constructed in the context of a far more complex geography of social relations, both internal to the region and stretching nationally and internationally, than any unifocal map, with London at its centre, could imply.

This reference to social relations stretching 'beyond the region' – 'nationally and internationally' – is significant. For, as was stressed in Chapter 2, the region cannot be conceptualized as closed. What we would add here to the basic argument of the last chapter, however, is a reflection on the complexity, strength and multifocality of these external relations. As indicated in the Introduction to this book, the south east as a whole is strongly connected internationally. What is more, the nature and direction of those international links are distinct from those found in other regions of the country. But this is to think of the south east 'as a whole'. In fact there is also considerable variation within the region. The City of London is a dominant node, its connections with other regions and with other countries and continents far outweighing in significance its more local connections to the region within which it is locationally set. It is a place which is internationally embedded. To describe the geographical location of the City in terms of the social relations which sustain it and give it its identity necessitates beginning at an international level. Moreover, this is a set of international links with a particular geography (they are links to particular parts of the world) and they constitute a network in which the City forms a node of considerable power and dominance (see, for instance, Amin and Thrift, 1992).

A similar point has already been made about Cambridge. It, too, is crucially set within an international network of economic relations, this time with science and technology forming the content of the link. Indeed, the high-technology dynamic of growth is as a whole one which importantly functions within an international context. But it is a different international setting from that of the City – it is more restricted in the parts of the world which it reaches, its internal geography is different, and so are its most powerful locations. Moreover, within that international space, Cambridge, and other places within the south east where high technology is important, are not points of power and dominance. In this sense, too, then, this international geography is unlike that of the City.

These examples could be multiplied. We shall see below that even some of the 'holes' in the regional lattice of growth have their own international connections, which are different again. The point is that this indicates a level of complexity and multifocality which goes beyond that indicated by Garreau. The 'openness' of the south east itself bears witness to the complexity of its internal structure.

There are, moreover, other aspects of internal complexity which go beyond the arguments made by Garreau. By definition, for instance, since the concept is necessarily comparative, not everywhere in the region could be a hot spot.

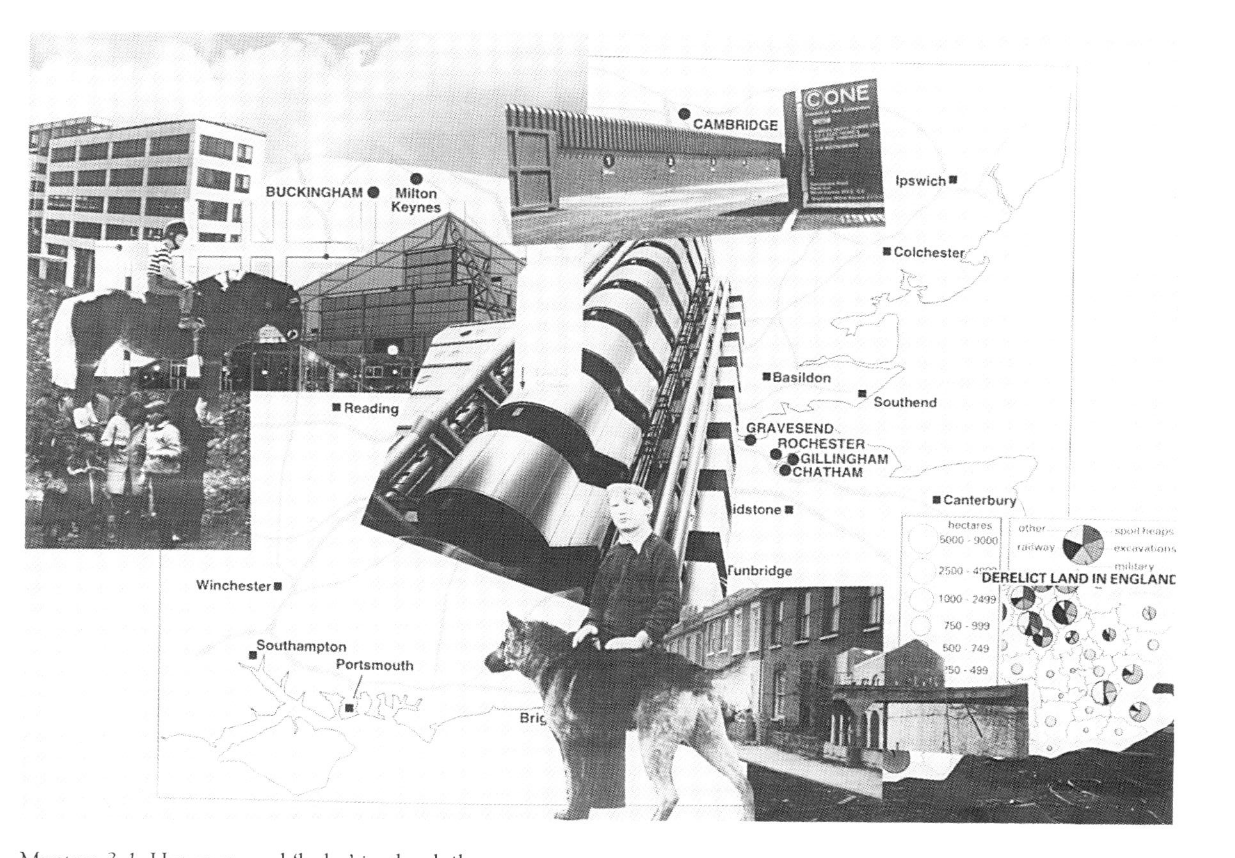

Montage 3.1 Hot spots and 'holes' in the doily

Clearly, there are distinctions between places within the south east, and not all of the south east fits the dynamic image of being able to choose between welcoming and resisting growth. In some places, the presence of growth mechanisms was more muted – there were fewer of them, with less in the way of mutually reinforcing effects perhaps, or individual mechanisms were less strongly present. In other places, the current round of growth was overlain on other processes of decline, thus reducing the net gains. In yet others, the growth mechanisms were barely present at all. We would argue that it is important to distinguish between these different cases and that our approach to growth is one way of doing so. In particular it warns against any conceptualization simply in terms of differing *levels* of growth.

The first two kinds of cases are simple. A number of places within the region fell clearly within the second type: that is, where growth in new employment – usually in services – was overlain on substantial decline in other sectors. Thus, Luton continued to lose manufacturing employment but witnessed growth in its service sector, although this had been underrepresented at the start of the 1980s. Other examples include Hertfordshire, where, despite a recovery from job loss in the early 1980s and an image of a prosperous (even overheated) area, overall employment growth was surprisingly low compared with other parts of the south east. The gradual decline of the county's defence and aerospace industries and the later impact on the county's New Towns where defence firms were the major employers was compensated for by substantial growth in high-technology (non-manufacturing) sectors, and the relatively short-lived 'boom' in other new sectors and jobs, accompanied by a national rhetoric of growth and (local) complacency about the future. What we see in such places are the quite different temporalities and spatialities of distinct 'rounds of investment' – a now declining though relatively recent manufacturing Fordism, now on the retreat, and a sprouting of new, private sector, services.

But this distinction between the geographies of rounds of investment shows up most clearly in cases of the third type: those where the growth mechanisms of the current round were barely present at all. These are the 'holes' in the regional doily of which we spoke in the last chapter. It is the existence of these places that raises the possibility of a spatially discontinuous region of growth.

The Medway Towns in North Kent are a case in point. These towns – Rochester and Gillingham – were largely by-passed by the growth of the 1980s and in this contrast sharply with most of the rest of the spatially contiguous south east. The loss of thousands of jobs in the early 1980s from their dockyards, primary and manufacturing industries was regarded locally as having an impact both in the sense of a rise in unemployment and also a psychological impact, as the dockyards in particular had provided a major source of employment for generations of (male) workers in the medway towns and beyond. Local authority officers considered it, at the time, as the 'end of the Medway Towns'. In trying to construct a role for themselves, the medway towns local authorities have vigorously pursued urban and economic regeneration strategies, seeking partners in the private sector and working with other North Kent local authorities and the County Council. In recognition of sharing joint problems, the local authorities have largely appeared able to put aside parochial and political differences in pursuit of a common strategy. Perhaps in recognition of being located in, yet not really part of the south east, their promotional activities for attracting inward investment have also built on their location in a 'new' European region. However, in common with many other places, the ubiquitous heritage theme is also strong in the Medway Towns 'image' campaign, building on Charles Dickens' links and finding a new function for the Georgian dockyard (see Charlesworth and Cochrane, 1994).

In recognition of the scale of the problem, the area won Enterprise Zone status (with the exception of London, there were no other EZs in the south east) and became the only location in the south east to work with the central government development agency, English Estates. This partnership between English Estates, the local authorities, the private sector and a charitable trust drew up ambitious plans to redevelop part of the dockyard into a working museum, convert neighbouring Georgian buildings into luxury housing and build a new community on St Mary's Island – homes, shops and other facilties – together with commercial and leisure development. Unfortunately, Chatham Maritime has not proved to be the success anticipated: conceived at a time of growth and optimism and a desire to replicate London docklands (without its problems), a delay in starting caused by technical problems consequently caused it to be a victim of the late 1980s' downturn in the south east economy and housing market. The Historic Dockyard museum was completed on time but has not experienced the huge numbers of visitors predicted. Whether the area would have been successful if launched at the height of the 'boom' is debatable: building a new community in an area of former docks and barracks, declining manufacturing, perceived as the 'poor relation' of Kent and on the fringes of the south east is a task beset with problems. (In the 1990s, hopes for success have been revived once again with North Kent's involvement with the east Thames Gateway regeneration initiative.)

This kind of story raises a host of issues. One thread which stands out is the importance of 'image' in this whole process. Historical representations of places can have significant force. In local pockets of the south east, such as the Medway Towns, the difficulty of breaking free from persistent images of decline made some places seem almost non-starters in the effort to participate in the forging of the south east's new image. And this was not for want of trying. As the paragraphs above indicate, the local authorities in the Medway Towns made considerable efforts, including specific efforts at image-refurbishment, but to little avail.

Such areas did not simply 'miss out on' the growth taking place around them. This is not simply a case of a kind of geographical trickle-down effect failing to reach the last bits of the region. On the contrary, the exclusion of such areas, including the Medway Towns but also many others, as indicated in Chapter 2, reflected the very nature of growth in the 1980s. The excluded areas varied in their character: some of them were places of persistent decline, others were areas with social mixes in terms of either class or ethnicity deemed to be inimical to the kind of growth dominant in the neo-liberal round of investment. Either way, they had the 'wrong' preconditions – social or environmental, 'real' or sometimes more imagined – for the kind of growth which was on offer.

Every kind of growth, and every round of investment, will have particular requirements and will map out a distinctive geography. There has been much written in recent years about the footlooseness of capital and about spatial indifference but in fact the locators of investment retain highly sensitive geographical antennae. If, at the level of 'the south east as a whole' the 1980s' neo-liberal growth saw the piling of growth upon growth (very different, for instance, from the nationally more dispersed growth of the social democratic project of the 1960s), looking within the region reveals the nuances and the precisions in choices of location in the local geography of growth.

One of the characteristics of this form of growth is its sensitivity to social conditions. Certainly, there are elements of investment which seek out locations primarily on grounds of, for instance, lowest costs. This pursuit of cost-minimization – of the cost of labour, of land and buildings, of infrastructure – has indeed been one of the mechanisms propelling those elements of growth in the wider south east which have been characterized by decentralization from London. These are long-established location factors. But there are newer ones now, which appear to be increasingly important. And it is here that 'image' becomes of salient significance. A 'setting of success' can be a vital part of a company's location decision. This can take many forms. It can mean acquiring the right address, even the right post code. 'Cambridge', for instance, has long been notoriously sought after by those wishing to attach to themselves an aura of pioneering scientific endeavour. It can mean paying attention to local setting and architectural design. And it can mean – and this is the important point here – avoiding like the plague anything which might give off an image of decline. What we begin to see here, then, are some of the differentiations – local and quite precise, but very effective – in the social conditions necessary for recent forms of economic growth.

Now, as we have said, every form of growth will have its own, particular, appropriate social conditions. These conditions will vary, in other words, between rounds of investment. However, what is distinctive about recent forms of growth, and this is a characteristic not entirely particular to but significantly reinforced by recent growth of the neo-liberal variety, is this singular importance of image and social character on the one hand and the avoidance of decline on the other. The social and economic implications of this are serious. At its most basic it means that the places most desperately in need of new investment, places suffering from the decline of previous forms of growth, are precisely the ones which are avoided. There is a deeply ironic geography in which places which do not need new investment are easily capable of attracting it (and in some places, such as South Buckinghamshire, actively mobilize to deter it), while places trying their level best to get some new growth to compensate for old decline (such as the Medway Towns) find the struggle virtually impossible.

There are some other ironies in this, in so far as this kind of geography is characteristic of neo-liberal growth. For neo-liberal economic theory at the level of political rhetoric claims precisely the opposite. Either it urges those in areas of decline to hang on and wait for the geographical version of the trickle-down effect (and at different spatial scales such places are still 'waiting' – the Medway Towns within the south east; Liverpool, for instance, within the UK; sub-Saharan Africa on the global view). Or it is argued that there is a geographical version of the more general processes (hypothesized processes) of equilibration whereby the very lack of growth can be turned into an attraction. Thus, it is argued, in the end things even themselves out. What we see from the south east is that this is very far from being the truth of the matter, even within the core regions of this type of growth. The reproduction, even the exacerbation, of spatial inequality is a structural effect of free-market economic strategies.

This, then, is an important aspect of the reproduction of geographical uneven development under neo-liberalism. It has a strong class aspect, and it results in exclusion. What we see here, at a more detailed spatial scale, is a further aspect of that 'two nationism', discussed in Chapter 1, which in wider ways was endemic to the pursuit of this politico-economic project. What we also see, of course, is one aspect of that geography of polarization which was discussed in Chapter 2. The Medway Towns, Thanet, Sheerness, parts of London, are holes in the doily of the south east which represent the geographical face of the exclusionary nature of this form of growth. The question which remains – and which will be addressed in Chapter 5 – is whether the crucial point about this geography is the fact of its marked inequality or whether such exclusions may become fissures exposing a potential vulnerability in the model of growth itself.

The image of these excluded places could not be in starker contrast to that of the City of London in the 1980s. As a place of the intense articulation of a series of growth dynamics the City, along with its remade images of yuppies, money (as against finance), technology and international dominance, provided much of the hegemonic imaginary geography of the region as a whole. Indeed, along with Cambridge and its images of high-tech entrepreneurship, these two places together were highly significant (representationally) in redefining the region over this period. It is to these locations, and in pursuit of further arguments, that we now turn.

Globalization, masculinity, class and space

'The City' of London and the city of Cambridge have long been places of eminence. Images of 'the City gent' and of the scientific genius (Bacon, Newton, Hawking) have long fed our geographical imaginations of these localities. These places are integral to the British geography of status. Their very architectures – pompous banks and peaceful quadrangles, Threadneedle Street and the chapel at Kings – resonate with the settled hierarchies of the old English class system. In their different ways both these places survived until the mid-twentieth century as territories of the old, male, ruling elite. Indeed, they were not just territories in the sense of being the locations of these strata, they were veritable strongholds.

Those strongholds have now to a significant extent been challenged. It is not in any way that there has been a democratization – their elitism and their masculinity, as we shall see in the next chapter, remain largely intact – but there has been dramatic social and spatial upheaval: a breaching of the old walls, a reworking of the terms of dominance. Moreover, in both cases this shock to the system has stemmed from the re-articulation of these places into new sets of elite international relations. What has happened to them is symbolic – though in extreme form – of much of what has happened in the wider south east.

Historically, the City of London has been identified as *the* place of finance for the UK economy. Yet, by the end of the 1980s, the City of London was not what it used to be; it held a different sense of place. It was not different in the way that Cambridge was, where the 1980s' growth of high-technology industry transformed the economic base and identity of the town. Neither was it, like Milton Keynes for instance, driven by a desire to construct a place and an identity for itself in the economy. Rather, at the decade's end, the City was still the place of finance. Indeed, in many ways, the boom it experienced during the 1980s consolidated its pre-eminent position in the UK economy. Yet the boom was part of a much greater historical transformation of the social and spatial relations of the world-wide system of finance and, thus, the City now held a different sense of place because as the world system of finance was reconstituted so too was the place known as the City. For the City of London, spatial repositioning within a finance system which was itself undergoing dynamic change was the key to both growth and a changed sense of place. The constellation of historical dynamics, but with a particular intensity, reformed the interdependencies and relations of the City and its uniqueness of place. In essence, as was pointed out in Chapter 1, this international city went global (Pryke, 1991).

As a high-profile example of the reconstitution of a place and its people, the City spawned one of the enduring images of Thatcherism – the yuppie (see relatedly, Savage *et al.*, 1992). The new 'sexy greedy' (Thrift, Leyshon and Daniels, 1987) occupations of broker, analyst and market maker encased in shimmering towers of glass and steel embodied the economic success, power and glamour of the boom years as the City became the high table of the dynamic articulation of global finance and property. As highlighted in Chapter 2, the City landscape became a pillar of nation-wide 'imaginary geographies' whether it be of 'the south east and the rest', symbolic of a changed set of power structures and the target of IRA bombs because of this, or the subject of satire in Mike Leigh films portraying the residential influx of the City's professional classes (squidge and sausage) into, for example, Hammersmith and Fulham and the potent 'changing social relations' between old and new(comers) that this entailed.

The yuppie was one (powerful) example of a new set of representations of finance 'put in place' during the 1980s and the advent of 'Big Bang', representations that challenged, and changed, the social and spatial identity of the City. Big Bang, and the deregulation of the equities markets that it entailed, as noted in Chapter 1, proved symbolic in the historical transition of the City and its social relations from international to global status. Pryke (1991) has outlined how central to this change was an increase in pace and intensity to life in the City (a process similarly experienced in the competitive space–time of high-technology Cambridge):

> The break with the sterling era meant that the City was exposed to the rationale of a new system of finance. The City was to become the hub not of a culturally familiar, slow-paced, empire-oriented regime of trade finance, but of a new fast-moving capitalism in which the City itself was to become equally international. . . . [Concerning] the so-called 'Big Bang' in October 1986 . . . [this] date was also to be the final challenge to the self-regulated securities markets. In the face of a new time-space agenda set by technologically reinforced, globalising financial capital . . . this was a period for the City to rearticulate its social power.
>
> (Pryke, 1991: 210–11)

The micro-geography of this social change can be detected even within the City's buildings. The social spaces of the conversational dealers are on the upper floors – oak-panelled executive dining rooms, fine art and all – whilst a few floors below, but still a significant distance above-ground, the screen traders deal across banks of telephones and computer screens (and across global space–time). New social strata may have arrived, but the social distance between the two sets of workers is maintained. Socially it is much more than a few floors.

The City is now constructed through a very different set of spatial relations from those of two decades before: it has, as is well known, an outlook and position which now spans the globe. London, along with Tokyo and New York, has become one of the three regional financial centres which complete the global system of 24-hour trading. More pointedly, as the City has reorientated itself to a global outlook so its relationship with the UK regions, including the south east, has changed. The City continues to be *the* UK site of entry into international finance activities with the headquarters of the vast majority of all authorized banking institutions located there, as are nearly all the members of the British Merchant Banking and Securities Houses Association (Court and McDowell, 1993). However, in contrast, certain domestic financial activities, such as retail banking and insurance, have looked to relocate from the City, front office and all. This pushes further an earlier decentralization process whereby the City had acted as 'front office' to a series of decentralized 'back offices' located, initially, in Croydon and Reading and then further afield in the likes of Bristol and Cardiff. Thus, this global reorientation can be seen to have increased the 'distance' of the

City of London from the rest of the UK (including even the south east) whilst, at the same time, the City has moved 'closer' to Tokyo and New York. The City, as many commentators have pointed out, has virtually become an island of finance in a different, reworked, global trading system.

In contrast, the historical construction of the place called Cambridge has been dominated by the University. Its combination of architecturally stunning college buildings, dons in gowns and punting on the river has created a potent representation of academe which sets Cambridge apart as a 'university town'. Throughout its local history, sporadic panics about the clash between 'town and gown' have brought reminders that the dominant identity of Cambridge as a cloistered seat of learning has not been all-inclusive.

> Our backs and bridges, bills and bells,
> Our boats and bumps and bloods and blues,
> Our bedders, bulldogs and Bedells,
> Our chapels, colleges, canoes,
> Our dons and deans and duns and dues,
> Our friends from Hayti and Siam,
> Tinge with kaleidoscopic hues
> This ancient city by the Cam
>
> (From T.R. Glover's *Cambridge Retrospect*, 1943; quoted in Stubbings, 1991)

Into this world has erupted, over the last two decades, 'the Cambridge Phenomenon', a mushrooming growth of high-technology activity and entrepreneurship. If the yuppie and the barrow boy symbolized, in however overstated a fashion, the decade's changes in the City of London then the scientist-entrepreneur, on the frontiers of science in the morning, wheeler-dealing with financial backers after lunch, played the same role in Cambridge. In both cases, the images exaggerated the real changes, but they none the less caught a certain reality.

The story of the Cambridge Phenomenon is that of the young male high-tech entrepreneur, often backed by the locally based venture capitalist industry created during the City boom, using scientific knowledge in the private sphere to create private wealth. It is also the story of the privatization and fragmentation of the research institutes of Cambridge (Webster, 1989). So great – and so hyped – has been the growth of this phenomenon that it now figures as an alternative, and competing, view of Cambridge in the 1980s. The landmark buildings of the high-technology industries have joined the colleges on the tourist itinerary.

Indeed, the contestation and coexistence of these two identities may also be argued to have formed a hybrid identity for Cambridge. For, up until the 1970s, just as public science was part of academe in Cambridge so, in contrast, in the 1980s private science has been used to present a different representation of academe in line with neo-liberal thinking. The phenomenon in the 1980s is symbolic (and symbolic only, see Massey, Quintas and Wield, 1992) of the argument about the need to 'bring the ivory towers into production'. This was epitomized in a series of national newspaper articles, such as, 'A trinity of trading, growing and making' (Lloyd, 1986) and 'Where dishevelled dons are giving way to boffins with more BMWs than bikes' (Buxton, 1988), and a distinct representation of the *academic* entrepreneur turning ideas into money was formulated. Today, then, it may be argued that the historical identity of academic Cambridge coexists and to some extent fuses with that of high-technology Cambridge.

As in the City, this transformation has come about as part and parcel of a reorganization of the international relations within which Cambridge is set. As a seat of academic learning, Cambridge was already a focus of international standing within intellectual circuits. Today, it is also a node within a high-technology division of labour at a variety of spatial scales. Nationally, it is the R&D location for a set of UK companies whose production (if it exists) is likely to be found in 'Silicon Glen' or Wales. Significantly, the same pattern of UK high-technology uneven development is being replicated by an increasing level of foreign direct investment, whether it be from the US, Europe or Japan. Cambridge, in particular, has been the recipient of much of this overseas investment (in R&D) as its identity as a source of technological innovation has spread. In this network of international investment, Cambridge R&D itself is likely to be only one of many such research centres spread across the globe. Cambridge is likely to report to Tokyo or Palo Alto and, as such, the town has taken its place within the increasingly integrated set of global high-technology regions.

In both the City and Cambridge, this re-fixing within global networks, and the social changes which have been integral to it, have also been the motive force behind a dramatic refiguring of local, physical space. In the City, the Bank of England had for centuries defined the Square Mile; it had lent legitimacy to certain locations rather than others. This was an example of the material and social control of space by an old establishment institution – the nightly clearing system only operated within the spatial bounds defined in her wisdom by 'the Old Lady of Threadneedle Street' (the Bank of England) (see Pryke, 1991). Fifty miles away, in Cambridge, the University played a similar role. Through a mixture of extensive landownership and ancient influence it was effectively the planning authority for the city. In the 1960s, notoriously, Cambridge turned down an application from IBM. The location of such investment in the town was seen as being somehow vulgar; it was 'out of character' with the place, with – that is to say – the University's view of how that place should be (Massey, Quintas and Wield, 1992). These places, then, were both emblems and physical manifestations of the settled old patrician rule which was both symbolic of a certain kind of 'Englishness' and one of the long-term foundations for the dominance within the country of the south east broadly defined.

As indicated in Chapter 1, and as will be explored in more detail in the next chapter, Thatcherism disrupted this old settlement. Thatcherism certainly entailed a form of class politics, but it had little time for the class politics (that peculiarly English class politics) of deference, and as a result the spatial strongholds of this deferential system came under threat.

In the City, the physical boundaries of the Square Mile were breached in the face of globalization: a combination of the simple pressure for expansion and the arrival of parvenu foreign banks who did not appreciate the expectations of automatic subservience effectively broke the old spatial matrix (Pryke, 1991). The virtuous articulation of the growth dynamics of finance and property was able to expand the financial space of the City, with its new 'smart' high-tech steel and glass towers, as far as Broadgate and Docklands. In consequence, the old stranglehold exercised by the Bank, the Stock Exchange and the Baltic Exchange was seriously ruffled.

In Cambridge, the change was enacted – at least at first – through decisions of the old ruling power itself. The town had never become a centre of nineteenth-century industrial development and in 1954 the Holford Report had vowed to keep things that way. And yet, by the end of the 1960s, things were shifting. In 1969, the Mott Report argued for an acceptance of a limited growth of science-based industry. The hard line of utter refusal had begun to soften. The Report spotted changes taking place in industry, and in particular the increasingly separable roles of science and technology. Acknowledging the then largely accepted linear model of innovation – that new innovations start among scientists and then proceed through development and finally into production – it recognized that each of these stages is socially distinct, that each requires a different form of labour, and – crucially – that they could now be separated out from each other geographically. The Mott Report saw the possibility of having the first, innovative/socially elite stages in Cambridge. This might effectively combine in one location economic development and social acceptability. Innovation could take place in Cambridge, but a 'company can move its *production* units to other areas particularly when this requires a different form of labour' (*Cambridge University Reporter*, 1969: 374; our emphasis) (see Massey, Quintas and Wield, 1992).

This slight slackening of spatial control, along with the attempt to maintain social control, was to blossom into the Cambridge Phenomenon. Quite what the University thinks today of what was unleashed in 'their' town is hard to pin down. The University still has the stranglehold of ownership in the centre of town, and strict planning policies are still in place, but rings of expansion now surround the town. What is clear, however, is that high-tech industry accepted the invitation with considerable enthusiasm. A complex set of reasons lay behind this, among them the presence of the University itself. One of the most significant reasons, however, is that of the town's 'image' and social tone. It is an elite location, even the presence of the University is significant as a location factor more for its symbolic resonance than anything else (see ibid.). Moreover, there is no disturbing legacy of nineteenth-century industry, either physically in derelict factories or socially in an organized manual working class. These factors have been crucial and they serve to highlight the significance of image and social structure in the neo-liberal form of reproduction of uneven development. While the medway towns plead for more investment, the centre of Cambridge chokes with uncontainable growth.

The City and Cambridge, then; both considerably reworked in their social and spatial relations, the old orders challenged in both, yet both still without a doubt elite spaces: pinnacles of finance, property, academe and high technology. These are the hegemonic images of these places: how they are seen from the outside and how they are lived and experienced on the inside by their dominant social strata.

Yet there are other 'Cities' and other 'Cambridges'. Every place is lived and understood differently by different social groups, and in the City and Cambridge these differences are sharply marked. Historically, both are places of glaring inequality. It seems to complement their pre-industrial bases: the effortless elitism of the financiers and academics has always required servicing by the ranks of the lowly paid and yet the actual lives, the activity and the spatial practices, of these people are rarely noticed. When they do figure – as in the poem cited earlier – they are part of a romanticized landscape ('bedders, bulldogs') within which others (the real occupants) live. As the poem says '*Our* bedders, bulldogs and Beddells'.

In recent years, the reworking of the dominant faces of these places (finance, high-tech) has gone along with a reworking of the spaces, both local and global, of these other lives. The ranks of the poor have been in part restructured, caught up in other globalizations, drawn into servicing other functions, inhabiting the back-stairs and the below-stairs of other buildings. One aspect of what has taken place is the transformation from an almost pre-industrial service class into the new poor of neo-liberalism. The next chapter will look more closely at this social shift, but what is examined here is its relation to space. For this is the other face – and the other spatiality – of free-market polarization. The medway towns are the spatial expression of the exclusion of places from growth. But both the City and Cambridge were both the foci of growth – they were the hot spots. The poverty experienced today within them – and both remain places of startling inequality – is not a result of the absence of growth; it is integral to the form of growth itself (Massey and Allen, 1995).

Whatever the alternative representations of the City, they do none the less share a coding of the place as finance. This combination serves to hide another side to the remaking of the City in the 1980s. If Big Bang represented one form of (politically inspired) deregulation which changed the character of the City, then the 'downside' growth of the contract service industries in the City is also the product of a decade of deregulation of public sector markets and the wider labour market. For a number of multinational contract service companies, the City has become a premier site of demand for the provision of contract support services and is tantamount to a hidden growth dynamic, with its own set of spatial relations and its own social form. Indeed, its lack of visibility is part of what constructs this 'downside' of growth (see Allen and Henry, 1995).

Moreoever, this growth dynamic is as international as its more powerful City counterparts. In the first place, the cleaning companies, catering firms and security organizations can be found alongside the finance houses in any listing of service industry multinationals. Indeed, the major operators of the contract services have grown in part because of the expansion in City office space during the 1980s' boom and the ongoing trend towards 'externalization' amongst firms. Perhaps more significantly, however, the contract work-force is international in an altogether different sense of the word, for many of its members are migrants. Thus, the coexisting work-forces of City professionals and contract workers each experience international relations at work; where they differ is in how these relations are experienced. Employees of a cleaning multinational, for example, are hardly affected in their daily routine by the international character of the firms which employ them. In a labour-intensive service industry whose major companies are recognizable by their fragmentation across hundreds, even thousands of workplaces, actual contact with the firm is rarely more than that gained through contact with the workplace supervisor. As we shall see in the following chapter, 'internationalization' comes, for instance, through a group of Nigerian cleaners reluctant to share their workspace with an incoming Ghanaian or Afro-Caribbean cleaner. The contrast with the internationalization of the mainly white, finance professionals could not be clearer.

The fact that only one set of social relations, that of finance, is core to the dominant identity of the City is important. For it means that the coexistence of the work-forces of dealers and contract services in the same buildings of the City does not imply that they occupy the same spaces. It is at this juncture that the power of the dominant identity of the City can be highlighted. For the smart buildings of the City *are* finance, and whilst they create unique and specialist demands of the cleaners, caterers and guards of the City, the presence of these workers is barely registered. Indeed, registered may be too strong a word, for the contract workers of the City are largely unseen or unacknowledged. In some sense, the work they undertake places them in the City, yet remote from it. Paradoxically, for many contract workers their inclusion within the expanding private services sector has meant their further exclusion from the socio-spatial matrix of the City, with all ties to their previous City employers effectively severed. And because this changed City is primarily constructed through its relations to New York and Tokyo, the relationship of the dealing room to other service spaces within the building is now even more of secondary significance. As the dealers discuss finance with their colleagues and clients, whether with a tray on their lap and over the phone to New York or over the table in the executive dining room, nearby chefs will be cooking, butlers serving, toilets will be freshened and passes checked in a world apart. The spaces – and the times too – of these different workers are sharply differentiated; and at night they set off home for very different parts of the south east.

In Cambridge, too, behind the dominant structures of high technology and finance other social relations, with other geographies, weave the place together and maintain its position in the world (Crang and Martin, 1991). These relations, too, have been transformed and, importantly, are critical to the success of the place. As in the City, the contract service industries are a less acknowledged element of the town called Cambridge.

Historically, the university town of Cambridge has always been a place of the serviced and those who serve. The colleges and their common rooms and high tables are the representational scaffolding of traditional Cambridge academe, quietly serviced by the 'Backstairs Cambridge' (Barham, 1986; Stubbings, 1991) of bedders, porters, bulldogs and cleaners. Yet even here, the 1980s' dynamic of 'contracting out' has entered in, with many colleges employing the same contract service firms as those found in the City. For the contract service multinationals, the spaces of Cambridge and the City represent an increasingly predictable extension of the continually expanding demand in the 1980s for their services.

However, in contrast to the University, the Cambridge Phenomenon of high-technology growth has proved difficult to respond to for these multinationals. For the nature of this growth – especially in the entrepreneurial 1980s – is one of extreme fragmentation and fragility. For the big contract multinationals, large numbers of small firms (or establishments) undergoing a constant process of birth, growth, decline, death and rebirth represents a series of small and varied contracts in a state of constant flux. Such fragmented demand has proved difficult for them to 'combine' in a form which allows them to bring their size advantages to bear. Instead, in Cambridge the decade saw the growth of a local contract service industry.

Like the finance houses of the City, the workplaces of high-technology Cambridge articulate very different space–times. These are the specialized spaces of research and development, globally if narrowly connected (that is, their links are with a highly restricted set of other parts of the world), yet they are guarded, cleaned and serviced by people embedded in very different spatialities and temporalities. These worlds meet, too, in the villages around the town. Many a Cambridge scientist goes home at night to an 'old world' English village, and may indeed quite often work there during the day. From a refurbished schoolhouse or a former labourer's cottage stretch e-mail and Internet connections, not just to the workplace in Cambridge, but to Silicon Valley, to a colleague working on something similar in Australia. In this way, work can be brought in to the home and still link in to networks worldwide. In another part of the village, the spatialities may be very different: the relative localness of daily life, with young mothers responding to a call from a local contract service company to come and do a few hours cooking, cleaning, or waitressing. Not only are the 'work geographies' of these participants in Cambridge's growth quite different, so too are the geographies of their daily lives. Indeed, even the space–time relations of home and work are utterly distinct. The contract-cleaning worker and the high-tech scientist both work 'flexible' hours, but they involve very different types of flexibility (see Massey and Allen, 1995; Henry and Massey, 1995). In contract cleaning, the hours are irregular, based on short-term contracts and often involve work at unsocial times. The scientists also work unsocial and irregular hours, but this is the world where long hours may be interpreted as a sign of machismo or brilliance, where you work until the job is done, where peer-group pressure for 'presentism' is considerable. This is the very opposite of hourly-paid labour. Moreover, these different temporal regimes are accompanied by different spatial practices. The contract cleaner *goes out* to work; the spaces of home and paid work remain separate. The scientist, as well as having his (*sic*) office, will also have a study at home. The sphere of paid work invades the space of the domestic (though, significantly, the process does not work in the other direction – see Massey, 1995a). And, while the contract cleaner may well come home from her paid work to turn her hand to the unpaid work of cooking, cleaning and waitressing for her family, the scientist will more likely rely on the paid or unpaid labour of others .

Overwhelmingly, their ability to *be* scientists, in the workaholic culture which has been enabled – and encouraged – in Cambridge depends on the care and support provided by others. The commercial success of high-tech Cambridge, with its glamorous spatio-temporalities, depends on the more local spatial practices and the complementary flexibilities of low-paid workers and obliging partners. The socio-spatial relations of this form of growth tie this same village in multiple ways into the reconstituted place that is Cambridge.

There are other elements in this complexity which will be explored in the next chapter. The long hours and the 'flexible' working, the unremitting (though very varied) masculinity, the utter dependence on someone else to do the domestic labour; all these things too have been integral to the remaking of these places in the 1980s. Much of the hype which now surrounds these places draws heavily on a glamorized space–time – jetting and e-mailing around the world, 24-hour trading, working deep into the night on some knotty problem on the (supposed) frontiers of science. But these space–times, and the lifestyles and the economic growth which they support, and which were absolutely key to the economic dynamism of the south east in the 1980s, are supported by other space–times divided from these dominant ones by class, ethnicity and gender.

Othering and interdependence

Place-identities are complicated things. Not only are they persistently multiple, but also they are formed inextricably out of the wider relations in which the places are set. The identities of the City and of Cambridge depend heavily on their international interconnectedness. The characters of the local and the global, in contrast to the implication of the 'Russian doll' approach which we criticized in Chapter 2, are constituted together.

There may be other aspects to identity-formation, however. The character of a place may be most forcibly established by clear assertions of where it is *not*. Social and spatial 'othering' may go together. Spatial barriers may be erected to maintain a valued social character or tight boundaries drawn to register, and operationalize, a social divide (see, for instance, Sibley, 1995). Such processes of identity-formation-through-counterposition can happen at many scales: Canada is *not* the USA, the people of mid-Cheshire are *not* part of Manchester. And it is certainly true that such processes operate, and with a fair degree of complexity, within the south east itself. Such is the intricate case of London, South Buckinghamshire, Milton Keynes and Slough.

Both South Buckinghamshire and Milton Keynes have long-established relations with London. From the former run high-class lines of daily commuting (you drive down the motorway or leave the car at the local railway station). The latter was dreamt up and planned (in the 1960s) as a new city, to be a counter-magnet to the capital, a place for overspill but not to be simply a dormitory town. Over the decades, those relations have become more spatially dispersed. Milton Keynes, indeed, might form part of a Garreau case for the south east as 'edge city'. Certainly, it has a high degree of economic autonomy. Moreover, both places retain an element of self-identification which operates through contradistinction to London. In Milton Keynes, clerical workers and taxi drivers still enthuse over how very much better it is here than in the East End of London whence they came. South Buckinghamshire functions as a hideaway for the wealthy, safely distanced from the metropolis and the other areas of new economic development in the region where their money is made.

These two places are chalk and cheese. Milton Keynes was born out of 1960s' modernism at its height. A place, above all, of the skilled working class. The residents of Milton Keynes are seen as representative of a 'new' south east person. South Buckinghamshire is self-conscious, manicured, residential rurality. Here are acres of private land and large houses, tree-lined roads and nostalgia for an olde England that never was. It is yet another element of that comfortable elite which we met in the City and Cambridge. The juxtaposition of Milton Keynes and South Buckinghamshire is like having Essex and Surrey next door to each other, and in both there is a consciousness of the anomaly. For South Buckinghamshire, in particular, it is a juxtaposition which is crucial in the establishment of the identity of the place. Milton Keynes, its neighbour within the same region of 'the south east', functions as its most significant other (see Charlesworth and Cochrane, 1994, 1997).

We have seen how some places in the south east have derived new dominant representations resulting from the articulation of growth dynamics and relations within and beyond the region. Yet in the wake of the growth in the south east, we can also identify places where dominant social groups have attempted to resist increases or changes in forms of growth, particularly the physical consequences of growth, such as new office and retail developments. Such development pressures have particular connotations for the south east, where there is a large amount of 'green belt' land. There are conflicts over balancing the need for jobs with retaining the quality of the environment. However, there are also examples of places where dominant social groups are able to exert considerable influence on planning issues, thereby preserving their perceived quality of life, yet at the same time benefiting from selected aspects of growth (high-income jobs in London, high house prices). Indeed, such places precisely reflect the importance of the consumption dynamic. Despite the more general lack of importance in the region of local authority boundaries in the construction of place identities, they are used by coalitions of local political actors and residents to protect 'their places' from invasion by the wider south east.

Yet, if the whole region is experiencing pressures for development, how can one small part of it be successful in resisting growth? We have already discussed in earlier chapters how the identity of the south east can be partly understood through placing it in the context of its relationships with other regions: the same is true for places *within* the region. In this example of Buckinghamshire, a more traditional image of the region is played upon, harking back to a time before high-tech connections to the rest of the world; in other words, a representation of the Home Counties as places to be preserved and traditions to be maintained (see Murdoch and Marsden, 1994 for a discussion of the role of planning in this process of 'reconstituting rurality in an increasingly urbanised world'). These are inevitably the values of particular social groups. The success of areas in resisting growth, therefore, might be explained partially by their relationship with other places in the south east which have, in contrast, welcomed growth.

Milton Keynes and South Buckinghamshire displayed stereotypical yet contrasting images of places in the south east during the 1980s. Milton Keynes fits the image of the ultimate 'edge city' – a potentially self-contained and expanding urban area, providing employment, shopping, leisure, housing and fast roads. In contrast, south buckinghamshire provides homes for commuters and is keen to cling to its perceived identity as a rural haven from London and its expanding suburbs. In South Buckinghamshire, the dominant group's focus is on how development affects their environment and quality of life and the place is largely introspective. Yet at the same time, South Buckinghamshire needs the continued growth and wealth-creation of the south east to preserve this way of life. For Milton Keynes, the situation is very different; it needs to look beyond the region to sustain its rate of growth and thus survive; hence, considerable emphasis has been placed on creating an external image which will attract investment and population. This has at times been influenced by the wider political context of the 1980s and by the types of growth heralded as essential to continued success, for example high-tech industries, service sector employment, foreign inward investment, private sector housing, etc. One thing the two places most certainly have in common is a conscious process of identity-construction.

Although it incorporated three existing towns in North Buckinghamshire, the architects of the Master Plan for Milton Keynes envisaged a new identity for their city, and certainly expected it to be viewed as a more ambitious project than earlier New Towns in neighbouring counties. In particular, images of American cities were at the forefront of the planners' minds. There was an emphasis on car ownership, a grid-road system, (enclosed) shopping mall and low-rise development. Milton Keynes was certainly constructed in the physical sense of new buildings and roads, but its subsequent identity was also 'constructed' – with a particular image in mind – largely through the actions of the Development Corporation.

The Development Corporation was regarded as innovative in that it incorporated both public and private sector actors with an emphasis on 'partnership' even before that concept was extolled by the Conservative government during the 1980s. However, the achievement of Milton Keynes's current and highly promoted image as a private sector success story depended upon an earlier period of huge amounts of state investment. In the early years, the Development Corporation built most of the factories, roads, houses, etc. and brought in the population. It was only in the early 1980s that substantial private sector investment in construction was attracted to the city. Arguably, this would probably have been the next stage in the development of any new town but it was given considerable impetus by the Conservative government:

> One of few things that Mrs Thatcher did that I admire was to cause a political shift in the attitudes of organisations like the Development Corporation . . . she had nothing to do with it directly but she created an atmosphere which caused that shift.
>
> (Interview: former Development Corporation employee)

The Development Corporation was encouraged to become more of a 'facilitator', working more in partnership with the private sector and eventually handing over to developers all 'large development opportunities' in anticipation of its own demise. This emphasis on the role of the private sector required a major change in the mentality and attitudes which had been prevalent in the city's early years: initially a public sector success story, it was now expected to become a private sector success story, locked into the 'enterprise' and growth rhetoric of the 1980s. Elements of the landscape of Milton Keynes reflect this shift during the 1980s: although often just separated by the grid roads cross-cutting Milton Keynes, some of the early public sector housing estates and the more luxurious 1980s' private sector developments are, socially and economically, poles apart.

The external marketing of Milton Keynes has traditionally focused on a number of issues: spaciousness, lack of traffic congestion, quality of life, housing choice (size, style, tenure and price relative to the rest of the south east), employment opportunities (and in targeting business – the availability of a skilled labour supply), leisure facilities, parallels with American cities (positive images only) and, above all, the benefits of modern living. For the incoming firm, Milton Keynes is presented as home to foreign investment, especially Japanese, high-tech firms, and a thriving service sector economy. To dispel its early public sector housing image, more recent marketing campaigns have targeted high income groups, adding currency to the promotion of Milton Keynes as a home for managerial and professional classes. It is an attempt to escape a 'down-market' image. In reporting these issues, local newspapers assume these people to be men (with non-working wives).

Furthermore, the imaginary geography of Milton Keynes is one in which, despite being planned and new, many of the settlements can still attain their own sense of place and community and apparently also experience the history and traditions of the local area. In line with Morley and Robins's (1995) discussion, the image-creators of Milton Keynes have engaged fully in the discourse of place-marketing in bringing together the perceived advantages of the area – making it seem more 'special' than other areas – together with celebrating heritage and urbanity. Therefore, not only can the citizens of Milton Keynes enjoy the benefits of living in an American-style city, they can also experience an English rural past, living in a 'quilt of secluded, but connected villages' (Bendixson and Platt, 1992: 173). Zukin (1991: 20) suggests that local image-makers construct either 'a microcosm of the past or a panorama of the future'; in Milton Keynes, you can have both (see also Charlesworth and Cochrane, 1997).

Despite falling in line with many image-creators' turn to emphasizing more cultural factors, the overall philosophy of Milton Keynes' key actors is still tied into seeking growth and economic well-being. The success of Milton Keynes depends on its continued growth. And this in turn has been, and continues to be, related to the success of the south east. During the 1980s, the consensual politics of growth in the new city continued (despite the end of the Development Corporation) with the flourishing ethos of partnership helping the presentation of a unified and shared agenda for growth. This, at least, is the image presented. Yet behind the public face of partnership, the carving up of Development Corporation responsibilities created new, and enhanced existing, tensions between the Development Corporation, the Commission for New Towns, the local authority, other public and private sector representatives and central government. In particular, the government appeared keen to prevent the borough council from regarding itself as the natural successor to the Development Corporation.

To what extent has the image of Milton Keynes pervaded the imaginations of people outside the city? They have probably become very familiar with the TV-advertisement images depicting happy children with balloons, traffic-free roads and an almost futuristic city centre juxtaposed with rural peaceful surroundings. Despite this, it is clear that for many who have never visited the city (and also those who have) Milton Keynes appears soulless, lacking in character and a sense of community, with all its housing estates identical. It may well be that, despite its attempts, the city's external dominant identity is worse than being viewed as a 'non-place place' – it represents the antithesis of the English Home County suburb. This is felt particularly strongly by its neighbours in Buckinghamshire:

> 'We're a bit worried about Milton Keynes' confided one of the Buckingham church ladies. 'It's getting a bit too close.' Her alarm was partly informed by the new city's spreading residential hinterland, but mostly by a deeper, more spiritual dread of modernity on the march. Fifteen minutes by road bridges, centuries in cultural terms. On the other hand, a triumphal young conurbation of glass and steel built for the motor car age: close your ears to the accents and you could be in Minneapolis.
>
> (*Sunday Times* 16 May 1993)

Despite their relief that somewhere like Milton Keynes exists to mop up growth pressures thereby preserving their rural or market town identities, the population and planners of nearby towns wish it were just that bit further away. This image of an expanding place (in terms of both development and the perceived values held by Milton Keynes residents) could also be considered a persistent one outside the county: despite its attempts to invent individuality, community and place, in effect it represents a mass market, still has relatively large amounts of public sector housing (and shared ownership with the local authority and housing associations), has a higher proportion of people employed in manufacturing than the south east as a whole, much of its population suffers social and economic inequalities on segregated housing estates, and there is a relatively high crime rate (see Cochrane, Seavers and Sarre, 1996). Despite all the rhetoric and advertising, one of its main advantages as perceived by people moving in – in the light of rising property prices during the 1980s – was cheap housing. Despite its centrality to the region in the sense of growth, conceptually it seems to remain on the periphery of the south east. Moving to Milton Keynes – socially – is a bit like 'moving north'.

Paradoxically, of course, it is precisely because Milton Keynes exists that places in the south east like Buckingham are able to sustain such representations as sleepy small-town rural idylls, home to Tory voters and traditional (and noticeably white) 'Little England' values. They can protest against development in their own environs because they know Milton Keynes is there to soak up development pressures. The south of Buckinghamshire, in particular, has exploited the existence of Milton Keynes as a 'legitimate' reason for rejecting proposals for development:

> [Buckinghamshire] is . . . a part of the south east where a conscious bargain was struck – no development down there in return for development in Milton Keynes. . . . There's no love lost between us [Milton Keynes and South Buckinghamshire] . . . if I was living down there, I wouldn't want any more growth and feel entitled not to have it because that's what Milton Keynes was for. . . . So whereas I get very irritated by Hampshire and other nice places which resist growth, curiously I feel that people in South Buckinghamshire have got some

justification because the County sacrificed its northern end as a growth point.

(Interview: local planning consultant)

The south of the county is comprised of two small local authority districts – Chiltern and South Buckinghamshire. Their identities revolve around small towns and villages, 'green belt' land, high house prices, a dominant middle class and a resistance to change, and especially to new development. South Buckinghamshire is the southernmost district and was created in 1974, effectively acting as a buffer for the south of the county, dividing it from Slough, which became part of Berkshire. The gulf, both material and symbolic, between white, middle-class, leafy South Buckinghamshire and more ethnically and class diverse, 'socialist', urban Slough is immense. This, coupled with South Buckinghamshire's location near major motorway networks and Heathrow, provides a powerful illustration of resistance to growth and the interdependence of places with opposing agendas in the face of the growth of the 1980s.

Although the perceived threats to their way of life in South Buckinghamshire have inevitably existed for some time, the national attention on the growth of the south east during the 1980s brought particular concerns to the district. It experienced increased pressure for growth and development. There were proposals for large retail sites, office headquarters, housing, leisure facilities. Most of them ('thankfully') were successfully resisted. Only small-scale redevelopment has been permitted and, even then, there have been attempts to restrict employment – since more workers would herald increased traffic and demand for housing in the district. The issue of providing social housing has been discussed, with one proposed solution to this 'problem' being to locate it (and thus the South Buckinghamshire residents requiring it) outside the district.

Although South Buckinghamshire has traditionally pursued an anti-growth stance, the strength of this was inevitably tested during the 1980s. The existence of Milton Keynes and South Buckinghamshire's 'green belt' location provide a strong explanation for the continued success of the resistance. The local authority has a strong Conservative majority with token 'opposition' from a small number of Independents and, on very rare occasions, conflict with the planning officers. However, there is an added dimension which may particularly distinguish it from other similar 'green belt' districts – South Buckinghamshire's dominant social class. These residents are clearly recognized by the local authority as a powerful lobbying force. There are various societies which work through extensive letter-writing on planning issues, attendance at meetings, contacting councillors and local newspapers, and which also operate indirectly as well, by having contacts in government or key civil service departments. One local authority officer summed up his opinions of the local population:

there's so much money around and articulate middle class people who have a lot of influence and clout . . . it's amazing the number of

> celebrities and wealthy company people who live in this area. They've got a nice little patch for themselves and to hell with everyone else. People have got lovely plots and big houses, nice cricket club, and they don't want other people to get in. The only thing that has spoilt it are the motorways because they didn't have any control over those. . . . If someone wanted to build a supermarket in Beaconsfield, I imagine they'd go mad.
>
> (Interview: local authority officer)

From the spatial to the social

This chapter has focused on the detailed geography of social relations. The individual places within, and the internal structure of, the south east have been conceptualized from the same perspective as was the region as a whole in Chapter 2. These smaller localities too are understood as the product of the spatial articulation of social relations. And articulation as it is defined here reinforces the image of a discontinuous region which is comprised of places of internal variability, unboundedness and porosity. Moreover, reading the region this way enables a clear entry into questions of the construction of place identity. Connections with other places, whether positive networks of linkages or the distancing gestures of counterposition, are an important aspect of the meaning given to places. To see 'the south east', or indeed any region, this way is to see it as a highly flexible, open, complexity.

This complex geography of social relations was reworked in the 1980s. The organizing focus of London continued its longer decline. The complexities of, and differentiations between, internationalizations increased. Most of all the region took on new internal shape as the dynamics of neo-liberal growth took hold and expanded. It was, however, a growth with a very particular geography – even within this region which was one of its heartlands. The dynamics of its uneven development depended heavily on image and social structure, with the inevitable result of reinforcing geographical inequality. Some areas were by-passed, structurally excluded by the operation of this particular form of uneven development. Others – our hot spots – saw growth piled upon growth. Yet even within them another geography of inequality was secreted, a geography distinguishing between parts of towns or villages, a micro-geography of daily lives within buildings. These geographies were the spatial expression of the inequalities produced at the heart of the neo-liberal, 'free market', project itself.

4

SPACES OF IDENTITY

From the previous chapter, we have seen how places within the 'south east' are positioned through their relationships to other places, both near and far, and indeed draw their identity through such relationships. In this chapter, in a similar vein, we show how the identity of social groups is formed through such relational constructs, and explore the open-ended and contingent character of group identities in the context of the neo-liberal spaces of the south east. In particular, we show how the project of neo-liberalism accentuated certain social qualities at the expense of others, and in so doing 'pulled out', as it were, different sides to people. So, for example, in the 'new' south east, if we take the identities of the business entrepreneur or the financial dealer as icons, they are iconic only in the sense that they personify 'enterprising' qualities which are dominant within the imagery of neo-liberalism. Put another way, the political project of neo-liberalism, like any political project, possessed the capacity to 'make up' people and social groups in different ways by 'speaking' to qualities that were either latent within groups or already emergent. Both materially, through policy and strategy, and discursively, neo-liberalism legitimized certain values and qualities designed to have an impact on workplace and consumption practices, especially in the sphere of homeownership. How such political messages are translated by groups or adapted to suit particular ends, and how successful such groups are at mobilizing around those messages, is none the less a conditional process. Context is all, and in the particular space–time of the 1980s' south east, such cultural translations were played out in varying locations and under particular conditions.

By way of contrast, for those groups excluded by the political messages of neo-liberalism – the Fordist collective worker or the black ethnic groups which lay outside of the English 'nation' – the repression of their associated qualities offered them only limited scope for cultural translation in restricted contexts. For those restricted to the 'mass of bad jobs' available, in particular the low-paid, insecure service jobs which continue to expand, the challenge is often one of defining themselves into a new 'service' identity, with all the adjustment and negotiation required to find a place in south east civil society. Likewise, for the range of ethnic groups which fall outside of any representation of 'white Englishness'

– not only black British and new commonwealth migrants, but also non-commonwealth migrants from Spain, Portugal, Colombia, the Philippines – the challenge is to construct an ethnic niche for themselves in locations culturally distant from the ethnicity of the white ROSE (rest of the south east).

Before we look at the spaces of exclusion, however, we turn our attention to the spaces created by neo-liberalism which opened up the possibility of new middle-class groupings taking shape in direct response to the culture of the established southern middle class.

Spaces of neo-liberalism

In the previous chapter, many of the hallmarks of neo-liberalism were played out across the region, with the spatial practices of neo-liberalism – its inscribed codes of risk and enterprise – both linking and setting places apart. Locations such as the City of London and Cambridge, among others, represented the sites at which the cultural response to the established middle class and its network of privilege were frequently experienced in their sharpest form. As locations of establishment power, defined in part by a professional elitism and assumed privilege, the 'gentlemen' of the City and the 'dons' of Cambridge epitomized 'one nation' Conservatism and thus an obvious point of difference from the practices of neo-liberalism. Defining themselves in relation to, but not wholeheartedly opposed to, the established characteristics of a southern middle class, the 'new' middle-class groups forming in the region were not just different, however; their very identity was dependent on successfully imposing a particular definition of 'establishedness' on the former group. As such, the combination of elements which forged the identity of the new middle class from the neo-liberal colours of enterprise, individualism and competition was never fully closed. It remained open and contingent; open to reinterpretation and to subversion by those resisting imposed definitions, and thus contingent upon the very social forces it wished to displace. All social identities, following Laclau (1990), may in this sense be considered as *dislocated*, precisely because they are constructed in relation to varying others. In the process of identity-construction, the ambiguous combination of elements which make up a grouping draws upon both their interconnectedness to other groupings and their differences. The result is not a simple series of opposing identities, but rather a related, yet different and often contradictory set of social identities.

In what follows, we look at the ways in which the versatility of the neo-liberal project gave shape and direction to new middle-class identities, precisely through its ability to impose difference through definition – in this case primarily by disrupting particular masculine identities.

City players

At first sight, it may seem rather excessive to allow what is a rather small occupational grouping – the financial dealers of the City of London – to carry much of the weight of the 'new' middle-class response to the established conservative bloc. The context of this particular response, however, is revealing in ways which display not an absolute clash of cultural values, but an attempt to reinterpret the 'money business' in modern, neo-liberal terms.

As noted in Chapter 3, the challenge to the established City community in the 1970s and 1980s came about in part through the introduction of US practices and the increased foreign ownership of the City's financial institutions which neatly silhouetted the 'gentlemanly' ways of doing business in the City at that time. A faster, more open, competitive style of working stood in clear contrast to the rather staid, collusive style of operation which played itself out at a slower pace and rhythm. Encapsulated in Cain and Hopkins's (1986, 1987) cultural description as a form of 'gentlemanly capitalism', by the time the Big Bang reforms took effect in the late 1980s, it was relatively easy to isolate many of the outmoded, paternalistic practices in the City's markets and to identify those who performed them. Among the bank directors, treasury specialists, managers and directors of corporate finance divisions, the right public school, the Oxbridge ties, the correct accent and tailor, set this group apart, both as a privileged network and as an entity seen to be out of step with the new money markets.

It would be misleading, however, to portray this social grouping as simply out of touch with contemporary events and global changes. What is more significant here is the political and cultural context in which it was possible for such an established social group to be defined as a past, spent force. Sober paternalism is one thing, but it is quite another to assume that this established City elite was anything but enterprising and competitive in its own terms or averse to speculation and risk. Yet the challenge to this group's social identity, coming from those mobilizing around the cultural messages of neo-liberalism, only gained meaning from the portrayal of them as competitive in a rather staid way or enterprising in a cautions, stolid manner. As financial 'players', they lacked the aggressive, racey edge required to play in today's money markets. Or rather, they were defined as such by an emergent group of dealers and financiers organized around a different but related set of competitive codes and practices.

Related in the sense that the two groups of 'players', and here this could be extended to include other commercial practices in the City, draw their social identity from one another, as well as from the global networks of which they are a part. Certain characteristics of the established culture were positively valued by the emergent grouping, in particular their consumption practices, but other characteristics formed the başis of a negative evaluation. What, in part, distinguished one from the other may indeed be traced to the spread of 'American' working practices, but more significantly, the two groups drew much of their cultural identity from contrasting versions of *entrepreneurial masculinity*: one anchored in a set of meanings around paternalism and 'gentlemanly' conduct, the other locked into a more macho, aggressive masculinity (McDowell and Court, 1994b, 1994c). Where the former group may well have constituted itself in the heady days of Empire as a thrusting identity, in the current political context its identity has been re-positioned to one of staid masculinity. As an articulated set of elements, the social identity of the City elite has been largely disrupted by the emergence of a social grouping antagonistic to its cultural position.

During the 1980s, for instance, merchant banking and a range of City occupations represented the apotheosis of success, power, glamour and, above all, a certain type of masculinity. They were dubbed 'sexy greedy' by Thrift, Leyshon and Daniels (1987), who pointed out just how many of the City's high-status occupations – the brokers, analysts, market makers and sales persons – caught the popular imagination as the essence of a neo-liberal lifestyle. With greater opportunities available in the finance sector than hitherto, men from a wider range of class backgrounds were seen to valorize a macho masculinity of hard-nosed, aggressive behaviour, replete with sexualized 'horseplay' and ribald language (McDowell and Court, 1994b, 1995).

The dealing rooms in particular were the site of this openly macho, or laddish, masculinity. Doing successful deals may be the key to prestige, but the maintenance of the relationships required to execute good deals involved participation in a culture which bonded in a particular masculine way: kicking a paper football around the office, for example, or throwing the secretary's soft toys around and inviting sexist comment. In fact, the language used to describe the dealing process was itself central to the maintenance of the 'laddish' atmosphere; one suffused with sexual metaphor, whereby dealers talked openly about 'consummating a deal', 'getting a leg over' or the danger of 'being legged over'. Much of this atmosphere is indeed neatly captured by one female dealer's observation that 'the worse men behave, the faster they get promoted'.

Of course, such an obvious difference from the culture of 'gentlemanly capitalism' was not valorized by those who still benefited from the networks of that culture. They in turn resisted attempts to portray their 'assumed masculinity' in any way other than positive terms and strove to subvert the new found 'tough' financial atmosphere of the City by labelling the new entrants as aggressive or at worst, 'barrow boys'. Group closure of this kind, especially by a group whose social identity was in the process of being undermined, amounts to an open play of difference – be it along style lines, dress codes, behaviour cues or whatever. In one sense, the text of neo-liberalism provided many of the cues and codes in this situation, with its legitimization of a particular kind of entrepreneurialism – a racy, forceful type of masculine pace and pressure – and its dismissal of a more measured, paternalistic style. A brash dress code, an ostentatious consumer style or feigning football in the office are thus, in this context, merely different forms of cultural translation.

For women in finance and commerce, however, neither the old nor the new cultural practices in the City offered a way into the south east middle class. The trading culture, in particular, was frequently defined to exclude 'non-masculine' women. Yet, although the working culture created by a dominant masculinity was generally viewed as inimical to women, there were some who had successfully negotiated a role for themselves. These women recognized that they could not beat the men at their own game, but that they could exploit the underlying dynamics of the sales relationship to their own advantage. The women who had made a success of their careers in equities, for instance, had developed a deep understanding of their relationship with clients and were able to exploit the dominant attitudes among their colleagues.

> As a market maker you have to be quite aggressive and women are generally less aggressive, I would say. But you can sell by using different characteristics – personality, for example. Basically, the whole point of being a salesman [*sic*] is you want your client to feel special, you want to make him feel he's right on top of your sales list. You want to get under his skin, you want to find out what makes this guy tick so that he trusts you and you develop this relationship. It's all about developing relationships, and as I say there's different ways which you can do that. I can't take anyone to a strip show and I don't usually get tickets for rugby and football matches which a lot of people are after. But on the other hand you can use your personality and your charm, the fact that they might like to be taken out by a . . . a pleasant girl, erhm, you know, it's different, it's having a different sort of character. It's being able to offer something different and as a girl if you've actually got that, you might as well use it.
>
> (Interview: woman executive, equity sales)

So, in defining themselves in relation to the dominant masculinity, few of these women appeared to be intimidated by the competitive working environment. On the contrary, they appeared to glory in it, thoroughly enjoying the buzz of frenetic activity when it descended on the markets. Neither had they sought to become honorary men, a strategy identified by Acker in her work on women in professional roles (Acker, 1990). Rather, they adopted an alternative strategy; one which recognized the constraints imposed by the male dominated working environment and the difficulty of challenging it, yet which used the pervading ethos, including its attitude toward women, to their own career advantage. They did not deny their 'femininity', as did those seeking to become an honorary male, but rather used it to empower themselves in the workplace and to compete against male colleagues in the struggle for success in the markets. Put another way, they acknowledged that traditionally feminine traits could be used to generate a competitive advantage and they had sufficient confidence in their ability to manage or control the process, whilst not allowing their use of femininity to be turned against them or used to undermine them.

Thus, although many of these women in finance were not in a position to challenge the masculine working culture, they did use aspects of that culture to secure a position for themselves within it. Moreover, they did this by developing a way of being at work which rejected the honorary male approach, with its implicit acceptance of femaleness as inherently disempowering, and enthusiastically embraced the possibility that women can thrive in, indeed enjoy, the challenges of a highly competitive and aggressive environment. In so doing, they challenged the notion that women are somehow inherently incapable of working in such conditions – reinterpreting competition and aggression for women, in opposition to those seeking to make them purely male traits.

Having said that, even where women do manage to accommodate the dominant social relations in the finance industry and find a way of working successfully within them, they face marginalization as soon as they try to combine working with raising a family or other domestic obligations. It is almost as if women can be forgiven for not quite fitting in, as long as their femaleness is not manifested in motherhood. For many women, the long hours that are required as a demonstration of commitment in merchant banks make it extremely difficult to combine working with an active role as a mother or indeed other sets of interests.

> There's a sort of feeling that it's a good thing to work long hours – actually I think this is a male characteristic – they will often stay late even when there's not any reason why they should do . . . I think there's two reasons, maybe they think it impresses senior people and there's also a sort of machismo to it – 'I can work long hours at my job and still perform well; it doesn't affect me' attitude.
>
> (Interview: woman assistant director, corporate finance)

While this particular form of machismo is obviously not an exclusive characteristic of neo-liberal cultures, there is none the less something about neo-liberalism which 'pulled out' this side of people and legitimized such behaviour in the name of 'enterprise'. The contingent, open-ended nature of these identities, however, ensured that such a formation remained incomplete and a sign of insecurity in such transitional times. Moreover, such cultural forms were not restricted to the spaces of finance and commerce in the late 1980s and early 1990s. Another set of masculine cues and identities which, in part, can be traced to the predominance of a neo-liberal culture at that time is also to be found in the high-tech spaces of Cambridgeshire.

High-tech elites

In comparison to the contrast of cultures in the City of London in the late 1980s, the response to the academic establishment was more of a cultural displacement than a cultural disruption. The context is this instance is one of the relationships between two kinds of elite spaces, both of which draw their currency from the

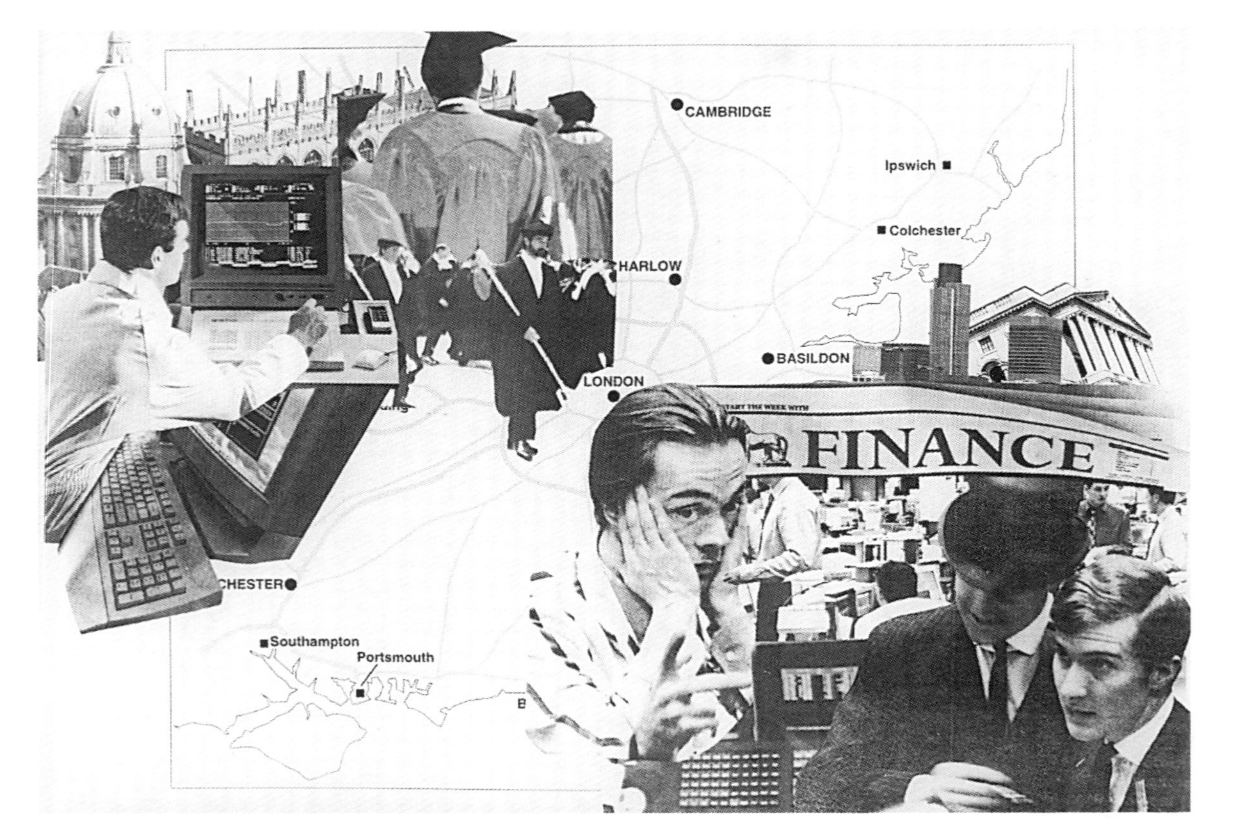

Montage 4.1 Entrepreneurial masculinities

fact that they are spaces of the mind, of knowledge and reason. One, as noted in the previous chapter, is represented by the 'ivory towers' of Oxbridge and the notion of the contemplative 'don', whereas the other takes its cue from the idea of knowledge as a marketable asset, something to be honed for commercial use in the rarefied atmosphere of the science parks. What distinguishes the two, therefore, or rather what enables the latter scientific elite to establish itself in contrast to the former scientific elite, is the very notion of *entrepreneurial knowledge*. It was not that the 'dons' were, in principle, averse to profit or commerce; rather the difference lies in what exactly was being marketed.

Expressed in this way, the characteristics of 'establishedness' which are undermined in this context revolve around the anti-enterprise culture of the academic institutions. Research and knowledge are not valorized for what they are in themselves, as pure, contemplative pursuits, but rather for their potential commercial application. The new figure, likened to that of an academic entrepreneur, was not only someone who could turn ideas into money; it was also someone who thrived on the neo-liberal climate of risk and competition. In a sense, the latter climate legitimized their activities and, as before, 'pulled out' the entrepreneurial side latent within scientific groupings. Deemed to be out of step on this occasion were those portrayed as either unable or unwilling to turn their research 'breakthroughs' into commercial successes. A label which, in this context, was pinned on the Cambridge groupings.

As we have had cause to note, however, the imposition of difference through definition is an ambiguous achievement, especially when the two class groupings share much in common. Both sets of scientific elites, for instance, would consider themselves as working at the frontiers of knowledge, struggling with puzzles of reason and logic. Moreover, it would be something of a caricature to suggest that academe in places like Cambridge was untouched by the marketplace or that competitive individualism was not a common feature of academic life – perhaps even more so today precisely because of the neo-liberal shift in values. Having said that, the broader context of the neo-liberal project did make it possible for a clearer line to be drawn between the two scientific communities as the entrepreneurial grouping grew in size.

Unlike the scientific thrust of the Wilsonian era, for instance, the commercialization of knowledge in neo-liberal terms rhetorically favoured the enterprising small firm as the vehicle for high-tech success. In contrast to a government-driven practice of scientific research and technology, whereby both public institutions and private corporations organized, for want of a better expression, large R&D factories, the neo-liberal initiative prioritized rhetorically and materially, the small, competitive business as the site of R&D. Linked globally through a network of contacts and contracts, the culture of competition and risk which permeates these new high-tech spaces is one that sets them apart from the traditional scientific community.

Within these high-tech work spaces, the interaction between scientists produces not only a culture which is 'work-rich', in Gorz's terms, but also a legitimization of pressure and competition which manifests itself in endless hours of work. Driven by the competition between companies in high-technology activities and the 'flexibility' of time required to meet deadlines and maintain customer responsiveness, the culture of 'ever-present' gives meaning to the identity of these scientific entrepreneurs (Henry and Massey, 1995). Like those in traditional academe, the scientists are 'always reaching higher', but on the science parks the new heights are commercially driven in a way that reflects the competitive structure of the high-technology business today.

Moreover, there is a further characteristic of this culture which shapes the identity of this set of middle class entrepreneurs, namely, its *form* of masculinity. At one level, there is nothing markedly new about the association of a culture of masculinity with the traits of reason and abstract truth. Whether it is scientists and their computers likened to big kids playing with expensive toys or academics taking delight in the solution to puzzle-solving logical games, the obsession with an abstract, instrumental form of rationality codes the performance as masculine (Massey, 1995a). In the spaces of high technology in the 1980s, however, these codings took on a specific neo-liberal hue.

Recall, for instance, the reference earlier to the long hours worked in the City of London's financial markets and their glorification as a sort of machismo. Among the scientist-engineers within high-tech firms, the pressure to work long hours was similarly a by-product of the masculine group culture, but here the drive was tantamount to a form of obsessionalism which mirrored the structure of commercial competition in the sector.

> We don't *need* to work longer – I think people choose to because they enjoy the work, because they own the project . . . and there's also ownership of the client.
>
> (Interview: male scientist-engineer)

Ownership of work, identification with the client, subjection to the pressures of 'not letting the side down'; each in their own way contribute towards an identification with a form of entrepreneurial knowledge which itself is reinforced by the organizational culture.

> But the thing we have discovered over the years is that people who work here, and get into it, become addicted . . . we find the problem of getting some people to leave; they do get very engrossed in the thing. . . . This circuit of people working on the system here, the difficulties are extracting them, for some other thing that may be necessary, like they haven't had any sleep for the last 40 years.
>
> (Interview: male company representative)

Sleep aside, there is also little in the rest of the lives of these scientist-engineers – family, children, domestic tasks and suchlike – that is allowed to displace their abstract work. For this reason, it is not altogether surprising that this occupational grouping 'lives' a deep split between work and home, with the latter excluded from the high-tech spaces of the mind (Massey, 1995a).

It is perhaps equally unsurprising to note that the overwhelming majority of Cambridge scientists are men, with few opportunities, in contrast to the City and finance, for women to be able to define themselves into the dominant form of masculinity or to use their femininity in such a way as to subvert the masculine codes to their own advantage. It is as if the more aggressive, 'laddish' masculine culture of the City is easier to subvert than that of the abstract, more instrumental masculine culture of high-tech spaces. Where one form of masculinity is open to negotiation, even if adapted or 'used' in a particular way, the other appears to remain rather more closed and defensive, making it all the more difficult for women to prise open a social space for themselves.

Indeed, in a general sense both forms of masculine identity were *constitutive* of neo-liberal growth across the south east in the late 1980s and early 1990s. They tended not only to mirror one side of Gorz's divide between the 'work rich' and the 'work poor', they also acted as a kind of boundary line for the inclusion or exclusion of certain social groupings from the neo-liberal project. A range of entrepreneurial groupings, for instance, shared the intoxication of the neo-liberal moment, yet each expressed it in their own particular way. Some, such as shareholders, represented a loose coalition of interests which none the less benefited from the neo-liberal moment, whilst others such as the financial dealers and high-tech traders discussed here formed a more coherent set of neo-liberal identities.

If it is reasonable to claim, however, that the neo-liberal project accentuated certain social qualities at the expense of others and in the process gave free reign to sides of people which might not otherwise have been fully expressed, it was the very 'success' of the south east as a growth region which fostered such transformations and which, in turn, derived strength from them. Above all, it could be argued that a particular kind of white enterprising masculinity came to the fore on the basis of the south east's 'success', legitimizing the activities of those groups which were capable of performing such a role and devalorizing the actions of those found wanting in this respect. Valorization, however, is not the same thing as aclaimed ascendancy and all such codes leave themselves open to reinterpretation and, of course, subversion. In fact, it could be said that the instability of the neo-liberal project in the 1990s, and the social identity of the range of groups associated with it, was not merely a side-effect of the political moment and its geography, but rather something inscribed within it. It is also true to say that what characteristics neo-liberalism excluded, or more correctly displaced, was the outcome of the very same political geography.

Spaces of exclusion

Exclusion from the neo-liberal project was not restricted to members of the established middle class. Other groups written out of the narrative of neo-liberal growth are those with the kinds of skill and experience which are considered inappropriate for the worlds of high-tech business and private services: the physical strength and industrial skills of the mass collective worker, for instance, or the bureaucratic skills of the redundant middle manager. In the case of either grouping, however, or for that matter any grouping displaced by the neo-liberal project, they have been directly exposed to the less predictable, more risk-laden environment of the competitive marketplace, although they face a rather different set of consequences from those groupings we have considered thus far.

The challenge for such displaced groups, as is often the case for women generally, is to define themselves into a context which is not of their own making; that is, to reinterpret themselves as part of growth, rather than be defined as a victim of growth. In this section, we look, first, at the experience of a group of mainly Fordist men who have struggled to maintain their identity and difference in an occupational space normally associated with women, namely, that of cleaning work, and contrast this with its mirror image in security work. Following that, we turn to the question of ethnic identity and to those groups directly excluded by the prestige of whiteness associated with the neo-liberal stress on 'race' and 'nation'.

Service ranks

The movement of men in greater numbers into jobs associated with women, such as cleaning and catering, is in part accounted for by the disappearance of jobs associated with an increasingly redundant form of masculinity: the kind of physical strength and skill required to work on a manufacturing line, for example, or simply the bland strength of the unskilled male. The expansion of employment opportunities at the bottom end of service labour markets across the south east, however, is more than the result of a sectoral shift to services; above all, it is the direct outcome of the liberalization of public services and of the broad shift towards contracting-out 'peripheral' activities which has accompanied the deregulation of the labour market (Deakin, 1992; Deakin and Wilkinson, 1992). Whereas the loss of Fordist jobs, for many semi-skilled and unskilled men at least, has primarily dislocated their class identity, the entry of this working-class grouping into the unregulated ranks of service work appears to have fractured their masculine identity along familiar and also less familiar lines (see Allen, 1997).

In much the same way, for instance, that a number of women in the City rejected the role of the honorary man as a way of defining themselves into finance work, so too working class men have resisted the role of honorary woman in the routine service sector by attempting to translate the work in familiar masculine terms – with varying degrees of success.

To take an example from the ranks of cleaning, whether part-time or full-time: a number of men employed in the expanding contract cleaning services have begun to reinterpret certain cleaning tasks, in particular those associated with machinery and 'real dirt' in line with longstanding masculine codes of industry. Women, for instance, have been operating various types of machinery in cleaning for a number of years, with little or no recourse to the characteristics of that machinery when describing the nature of their work or its imagery. With the noticeable increase in the numbers of men from traditional industrial sectors entering cleaning, however, the apparent 'heaviness' of certain machines and their 'speed' have begun to bring about a re-negotiation of tasks along gender lines. The larger the machine, the faster it spins or the more deft the strokes across the surface, the more likely it is that it will be operated by a man. Light hoovers are still operated by women (and also men), but the 'skill' of handling the more complex or weighty technologies is increasingly regarded as a job for men. In one NHS hospital in London, for example, the switch to contracting-out coincided with a gender shift in the tasks associated with particular types of machinery.

> We were domestics at the National Health, so it was all working machines and when private contractors came in, they were big and fast machines, so they preferred men on them. But I don't know why, because we can do it.
>
> (Interview: woman domestic cleaner)

In a sense, therefore, because more men have entered contract cleaning, a sexual division of labour is being *invented* in the occupation; one replete with skills that accrue largely to men rather than to women.

Relatedly, certain tasks associated with the use of cleaning machines are increasingly allocated to men, for example, the deep cleaning around kitchens or the buffing of floors with high-speed polishing machines. Although arbitrary in many respects, the legacy of industrial cleaning in manufacturing (that is, the cleaning of boilers or kilns which involved a combination of chemical technology and strength), provides the guiding text for where the gender line is negotiated in cleaning. Where men are part of a daytime 'team', even toilet cleaning has been added to the list of men's jobs.

This is not simply a case of men reinterpreting jobs in straightforward masculine terms, however; that is, drawing upon familiar masculine traits of strength and grime associated with industrial cleaning to legitimize a particular sexual division of labour. The division of tasks may be framed in such language, but the kind of 'identity work' which men moving into the ranks of cleaning have to confront is characterized by uncertainty and ambiguity (Thompson and McHugh, 1990). In drawing upon a stock of industrial masculinities, those men more familiar with manufacturing practices effectively translate an old script to meet the demands of a new work situation. There may be fewer industrial boilers to clean in today's industry, but the tasks and skills associated with such work are inserted into a new situation and invested with old meanings. Or, rather, an attempt is made by some to do so, which may or may not be legitimized.

There are other ways, too, in which men have attempted to code the work of cleaning in the masculine terms of identity. For example, although not yet widespread, there appears to be a greater willingness within some cleaning firms to construct full-time work for men. Much depends on whether or not men actually occupy the middle management and supervisory roles in a firm, but if they do there appears to be a greater readiness to identify a bundle of cleaning tasks to fill the 'working day' – which may well involve work on one or more contract sites – and to call it a 'whole job'. Some of these bundles are fashioned into supervisory roles (of which a number have always been full time), whereas others form the basis of a daytime cleaning 'team'. Once in place, however, strategies of exclusion come into play, often with the direct collusion of a male supervisor, to keep women out of the 'team'. Such strategies, although many and varied, often have a cultural note of masculinity – humour, wall posters and the like – which serve to erect informal workplace relationships around a set of shared values. Although such strategies are familiar on the shopfloor of manufacturing or in craft-based industries (Cockburn, 1983; Collinson and Hearn, 1994), their presence in occupations primarily associated with women is a more recent phenomenon.

To be clear, we are not suggesting that the bulk of men moving into contract cleaning are employed on a full-time basis. Many work part-time in the twilight hours before and after the 'working day' of others, but what is of interest in this respect is the frequency with which, so called 'post-Fordist' men seek to construct their own version of full-time work by taking up two or more part-time jobs to fill the day. In this case, only one element of their 'full-time' bundle is likely to be a part-time cleaning job. In contrast, women remain predominantly part-time cleaners; a state of affairs which embodies gender specific assumptions about both the nature of cleaning work and its timing.

However, in a challenge of this kind, which is constructed in relation to women's identity in service work, the ability of incoming men to define who is and who is not suitable to perform certain cleaning tasks and roles is strictly limited. Women occupy a range of managerial roles in the hierarchies which shape contract cleaning and their willingness to accommodate an invented sexual division of labour in the sector is likely to be minimal. As with all forms of identity constructed in relational terms, women managers have overturned definitions applied to them and to 'women's work' which did not correspond to their long-held views of who does what in the industry:

> we are out there all the time and we know what our girls are capable of and what the men are capable of, and all the staff. . . . We have girls that have been here 20 years and, more or less, you can leave them alone, because what is the sense of training a girl that has been cleaning for 20 years . . . at the end of the day, we want a very skilled person, who can clean properly. But I think everybody in the world can clean.
>
> (Interview: woman site manager)

And on men:

> We have men, I have got three wonderful men. For domestic use they are really good. We have about 36 area staff and two men. No, three now? No, four now? There is Saul, Caesar, Victor? The other one started yesterday . . .
>
> (Interview: woman supervisor)

If cleaning is one example of marginalized service work posing particular problems for certain men attempting to maintain their identity and difference in a new occupational context, another unregulated sector, that of security, has thrown up a different kind of challenge to male identity. Much vaunted in neo-liberal terms as a sector which operates on a competitive, free-market basis, security has long been associated with men and the masculine qualities of strength, vigilance and protectiveness. In such a context, the growth of numbers in the industry, especially across the south east (Allen and Henry, 1995), would appear, unlike cleaning, to pose few identity dilemmas for men entering the sector from mass manufacturing or the massed ranks of middle management. But this has not been the case.

The experience of insecure employment in the sector in comparison to most types of Fordist work is one obvious challenge faced by new entrants. Whilst there is little new about such an observation, what is different about security work over the last decade is the nature of the work itself, the skills required and, relatedly, the form of masculinity sought. As the market for contract guarding in the south east has shifted from manufacturing to commerce (offices, shops and such like) so too has the interpretation of what a security guard should be. In a real sense, the guard *is* the standardized product of the security industry. To be frank, nothing much ever happens in security work, apart, that is, from filling empty time and so there is little call to demonstrate actual physical strength. Whereas previously *he* would reflect the qualities of a strong physical presence or at least that of a 'warm body', today the emphasis is upon the presentational skills of guards (see Allen and Pryke, 1994). Under the guise of professionalism, the security guard in the City of London, for example, or at a high-tech plant should be alert, prepared, smart, ordered, responsive. Part of the presentation is that *he* should be both deferential and assertive – but not aggressive.

> It is very broad, you can be doing reception duties or you can be doing access patrol, you can be searching people, you can be doing the telephone or you can be dealing with customers, booking people in and out, you can be dealing with problem truck drivers, or executive visitors from elsewhere, so it's 'horses for courses'. When we recruit, we recruit specifically for whatever assignment it is, and the calibre of the person has to reflect that, so they sometimes have to be personable and able to reflect that and stand at reception and be well dressed, well presented,

> intelligent enough to identify some instruction requirements. . . . In a service industry, you have to deal with the customer and the contractor and the guards will always have at the back of their mind that if they screw up, we lose the contract, then they are out of a job, so there is that element as well.
>
> (Interview: male personnel director)

Interestingly, therefore, the emphasis is upon social skills, rather than craft- or technology-related skills. Despite the fact that teams of guards at large office sites are invariably surrounded by banks of security screens reflecting state-of-the-art technology, the identity of the occupation is rarely defined in relation to technology or 'machines'. In other words, men do not, as a matter of course, assign a technical skill to their work which may then be used to exclude women from this particular kind of security work. Nor, somewhat surprisingly, is the threat of danger implicit within security work often used to justify the job as 'men's work'.

Putting all this together, the outcome is a shift in the definition of masculinity within the security industry that places less emphasis upon physical presence and more on social presence. Obviously, not all men within security, especially those restricted to night work, fall within this definition, but this lack of correspondence does not undermine the shift in gender representation which has taken place in the industry. Indeed, the more physical form of masculinity has been subordinated by the construction of a more interactive, responsive, alert masculinity. Such characteristics, however, are less central to security work roles that are routinely gendered as male. As a consequence, there is less that defines women *out* of security work – especially given the fact that neither technology/machines nor 'danger' are routinely drawn upon to exclude – except the long hours.

> In security, it has got to be alertness really. I don't mean it's like an army, but you jump if you see something, you report it straight away, have your eyes open at all times, and notice things maybe he or she has missed. . . . Basically, the job boils down to remember this and go and get on with it. It's not like it's just a man's job, it can be done both ways. But why women are not in it? I think it is the hours really, seven days a week, twelve hours a day.
>
> (Interview: male site supervisor)

As with high-tech work and finance in the City, the issue of long hours is imbued with a particular form of masculinity. Less a form of obsessionalism as in the case of high-tech, or indeed an outward expression of a type of machismo as among many of the City's dealers, the long hours of security, especially the fact of night work, mainly reflect certain gender specific assumptions about those who can fill them. Quite simply, it is usually only certain kinds of displaced men – the old, mass collective worker in particular – who are 'available' to fill empty time, regardless of any other commitments. In fact, aside from the question of hours, the identity of much of the security occupation now overlaps with certain forms of femininity, which requires men to define themselves differently in masculine terms if they wish to be part of this type of service growth. The coming to the fore of an interactive, responsive, alert masculinity within the security industry involves a set of qualities for which much manufacturing work has no precedent. The inability of some men to negotiate such a role simply labels them as victims of growth and opens up more opportunities for women to be both the same and different from men in the job (Cockburn, 1991).

White city

The displacement of social groupings by the political moment of neo-liberalism – whether at the bottom or at the top of the social scale – did not only work itself out along class and gender lines, however. Ethnicity and 'race' had a central part to play, too, as suggested in Chapter 1 and intimated in earlier sections. The celebration of a different set of middle-class identities from those of the 'established' middle class did not alter in any significant way the whiteness of the ROSE or indeed that of the City of London itself. True, the whiteness of the ROSE was no longer overtly acclaimed as an Oxbridge whiteness or a genteel whiteness, with an unambiguous set of gender codes, but, for all that, under Thatcherism, the prestige of a white Englishness stabilized the culture of neo-liberalism and graphically illustrated who was 'out of place'.

As Hall (1990, 1991) has observed about English ethnicity, one of the reasons that it is hegemonic is precisely because it does not represent itself as an ethnicity at all. This is particularly so for the white Englishness of the City of London and indeed for the outer south east. They simply *are* English; but of a particular kind – one that, under Thatcherism, was locked into nation and 'race'. Those challenging the old white establishment also quickly took on their rural symbolism – the waxed green jackets and countryside accessories – and reinterpreted them to suit their own ends (Thrift, 1989). The power of this representation is that other constructions of Englishness, for example, black Englishness, are defined in response to it and, indeed, other ethnicities located in London have to negotiate their social identity in relation to it (Gilroy, 1987). In the City of London, for instance, a tiny percentage of middle and higher managerial jobs are held by members of ethnic minorities. In sharp contrast, at the other end of the City's labour markets, there is a mass of ethnicities to be found – not only black British and new Commonwealth migrants, but significantly non-Commonwealth migrants from Spain, Portugal, the Philippines, Colombia and a variety of other countries. At the risk of generalization, whereas white English men and women are most likely to seek work in low-paid services because of their lack of formal qualifications or the lack of demand for their skills, the different ethnic minorities to be found at the bottom end of service labour markets are there primarily because of their ethnicity – regardless of their skill capabilities. From the vantage-point of these groups, the city of low-status services is predominantly a 'non-white city', defined in relation to a white (sub)urban ethnicity (see Waldinger, 1992, on US cities).

Whilst the white/non-white representation conveys the major fault line of opportunity in the City of London, it none the less glosses over many of the ways in which different 'excluded' ethnicities have been able to define themselves into a particular form of growth. A broad understanding of the processes at work can be gleaned from the writings of Cross and Waldinger (1993), who have drawn attention to the emergence of distinctive ethnic niches in London. Differences in the characteristics of ethnic groups, such as qualifications, occupational predispositions, informal ethnic networks and other endowments, as well as the extent of discrimination, can be seen to come together in particular ways to channel groups into certain occupations. At a general level, their study shows the extent to which Afro-Caribbean minorities in London, for example, are increasingly found in the public services sector, albeit at lower levels than white English. Looking in more detail, however, it is possible to show how a similar process of ethnic distinction divides up jobs at the bottom end of service labour markets in global cities such as London. Once again, the example of insecure service work in cleaning is symptomatic of the processes at work.

One of the main characteristics of contract cleaning, or most contract service work for that matter, is that it takes place at someone else's workplace – the multitude of sites, be they banks, public buildings or commercial offices – which are scattered across a city like London. Many of these sites, however, represent spaces of survival for ethnic groupings, not for the larger Commonwealth groupings especially, but rather for the small groupings who are concentrated in central London: the Colombians, the Portuguese, the Nigerians, the Ghanaians or the Filipinos for example. Through an extensive pattern of ethnic networks, some of which reach far beyond the 'region', particular sites tend to have a disproportionate number of one ethnic grouping. In a real sense, an ethnic division of labour within cleaning work has sprung up in central London, with those who actually do the cleaning drawn in large part from a recent migrant background. The relative decrease in the number of white English women, as well as first-generation Commonwealth groupings, from this kind of low-status work in central London has (not for the first time) opened up recruitment to a whole new wave of migrants and ethnicities. With the outward spread in the number of black English groupings and the movement of the white (ex) Fordist work-force to new towns like Milton Keynes and beyond, different ethnic groupings have filled the occupational gap through their extensive networks – often in the process actively restricting access to groups other than their own.

> I think you notice it more with the managers, like we have a Portuguese manager, so all his staff are Portuguese. There is a guy who works in one of the hospitals, and most of his staff are Nigerian. . . . I've learnt a lot

> working with them. I mean at first I didn't notice it. I just thought it was them moaning, but it's right because if I go to a different contract and have to run it while someone is on holiday, you can see it then, in their staff. They are more prejudiced than the whites, I'll be honest, and I learnt that very very quickly. So you keep Ghanaians with Ghanaians, Nigerians with Nigerians, and work it that way.
>
> (Interview: woman site manager)

There is a twist here, however, in so far as the ability of such groups to exclude other ethnicities from a particular niche does not rest upon recognizable, well-established cultural differences between them. For such differences are constructed *in place*, as it were, in relation to other proximate identities. To be a Colombian in London, for instance, involves something qualitatively new for Colombian ethnicity, as meaning and identity is constructed in contrast to, say, 'black' Nigerians and 'white' Londoners. There is a translation process involved in which migrant ethnicities draw upon and rework their cultural identities in the context of London's and the south east's particular ethnicities, especially that of white Englishness (see Gilroy, 1993; Bhabha, 1994). The experience of being Colombian in London, therefore, is likely to be a fluid process precisely because there are no fixed 'others' through which a closed identity can be established.

Perhaps a less obvious point here is that understanding the variety of cues in this relational sense also illuminates the fluidity of white Englishness. If Hall (1990) is correct about the hegemonic conception of Englishness constructed under Thatcherism, then this too draws its account of 'race' and 'nation' from the specific time-frame of the 1980s and the varying others present in that space. Indeed, the very strength of the 'race' imagery associated with neo-liberalism can in some way be attributed to the proliferation of ethnicities marginalized by such a symbolic order. As Hall suggests, the strong positioning of 'race' and Englishness also produces 'a recognition that we all speak from a particular place, out of a particular history, out of a particular expression, a particular culture.' (ibid.: 58). The power of Thatcher's 'Englishness' however, as noted earlier, was that, like a number of dominant groupings, it did not represent itself as an ethnicity, and thus did not have to acknowledge that it spoke from the white culture of the (sub)urban south east.

The home front

If, thus far, we have shown that the reorganization of work and the impact of the neo-liberal project has produced new identities and insecurities for privileged employees as well as less privileged, such a condition is not restricted to the arena of work and employment. Housing has also been a site of identity-formation and contestation across the region. Increasingly, insecure employment, coupled with the 1980s' reorganization of housing provision, has had a dramatic effect on the ability of housing to exert an effect on experience and identity. Social identities are always the product of more than one dimension, and housing is part of that complex constitution – arguably a more significant dimension since the 1980s. Whereas, up to 1975, housing was becoming progressively more secure, in the 1980s and 1990s it began to have an independent, and often unpredictable effect on disposable income and wealth. As a result, it became a significant cause of inequality as well as personal and material insecurity.

For a century British housing markets had acted largely to consolidate success or failure at work. The Victorian market was essentially a free market for rented housing. Increases or decreases in income could quickly lead to moves to better or cheaper housing. The growth of owner occupation and council renting after 1919 created more secure tenure of better quality housing, but in each case access was rationed – by building society managers or housing departments. In these tenures, some appropriately qualified households (salary earners or skilled workers) could get access to good-quality housing at relatively low cost. In each case the benefits were experienced mainly as better-quality consumption, though owners were also aware of acquiring an asset which could be left to their children.

The notion that housing could affect identity was applied in Thatcherite policy and rhetoric in the 1980s, first in the attack on council housing and later in the direct and indirect promotion of owner occupation. The shift from renting to owning was expected to promote privatism, self-reliance and accumulation of an asset to leave to one's children. It could be presented as both contributing to financial independence and to positive family values. Towards the end of the decade, as house prices rose faster than incomes, these sober reasons were overtaken by fear of missing out: 'it's now or never' became a frequent reason for buying. House price inflation raised the stakes in the gamble on ownership: the struggle to manage the first years of a large loan became more testing, especially as building society deregulation had made mortgages more expensive as well as easier to get, but the future prospects for asset-accumulation looked even more alluring.

Through the 1980s there was a positive feedback process: rising council rents and easier access to mortgages persuaded more people to buy, raising prices and encouraging new building. Rising house values were widely discussed and increasingly stressed as a reason for owning. It appears that rising values also encouraged homeowners to spend money on consumption, thus accelerating the boom in the whole economy. By 1989, the Chancellor was obliged to raise interest rates to control the boom, and he also announced that he would limit tax relief to £30,000 per property rather than per taxpayer. Multi-earner households scrambled desperately to buy before the budget, but their problems were soon dwarfed by a collapse of prices and transactions.

During the first half of the 1990s, the owner occupied sector experienced its longest and deepest recession. Prices fell by a fifth, turnover fell from £2.15 million in 1988 to £1.13 million in 1992 (Hamnett and Seavers, 1995). A quarter of a million households have been repossessed by their bank or building society, often being left with large residual debts as well as becoming homeless. Over a million households found themselves in negative equity – having an outstanding mortgage larger than the value of their home. Negative equity was experienced most by recent buyers and was found to be most frequent in the south east and among low-income buyers (Dorling, 1993)

Hamnett and Seavers (1994b) have documented both the distribution of gains and losses by homeowners and their pattern of movement through the owner occupied market. In a random sample of owners in five areas in the south east, just under a half were in their first home, just over a quarter in their second, and 15 per cent in their third. Older households and those of higher social class were more likely to have owned several houses. First-time buyers tend to live in older and smaller, often terraced property, and the greatest improvement in quality occurs at the first move. Later moves seem to involve smaller steps in size and quality, though they may well involve moves out of the inner city into more prestigious locations. Eleven per cent of respondents admitted difficulties in keeping up payments, with higher proportions among recent buyers and those with lower incomes. Although proportions were lower, some long-term owners and some professional and managerial households were experiencing problems in keeping up payments.

Negative equity was still less a respector of persons: 42 per cent of households in negative equity had high incomes and 35 per cent were managers or professionals. The explanation lies with time of purchase: 74 per cent had bought since 1985 and 81 per cent were aged less than 39. Indeed, as the market approached its peak in 1988, only those with high incomes and good prospects could have afforded to buy, so there was an added risk to the better off.

Taken over the long term, the recession may seem a brief anomaly, but Hamnett and Seavers (1995) show that it dramatically reversed the steady asset gains from owner occupation which had been so lauded by authors like Saunders (1990). They detail a range of measures of changing value between purchase and the present, the best of which allows for the changing value of money and outgoings like deposit and cost of improvements. Even using a simpler measure, termed the 'real crude gain', which allows only for the price paid and any outstanding mortgage, they found that 58 per cent of households had made a loss on their current home and 40 per cent had made a loss over their whole period in owner occupation. Comparison with the British Household Panel Survey confirmed that 40 per cent had made a crude loss on their current home: 18 per cent of owners having lost £30,000 or more. BHPS has the advantage of covering all regions of Great Britain and shows that the south east had a higher proportion of losers than other regions – but also a higher proportion of high gainers. The roller-coaster of rising and falling values was far from affecting all households equally: the timing and location of purchases and sales could leave some households as significant gainers and others deep in debt. While longer periods of ownership and higher class and income were some protection, they were not complete safeguards. Disaster could and did strike the privileged as well as low-income buyers. Far from being 'as safe as houses', house purchase carries an element of risk: it now seems like a gamble rather than an investment.

So far, we have shown what actually happened to fragment the experience of homeowners, but have given no indication of how the owners felt about it or what they were trying to achieve. Saunders (1990) certainly argued that homeowners were moving for investment reasons in the 1980s, but no other researchers have confirmed his view. Sarre and Seavers (1995) present a quite different picture. When asked why they had become owner occupiers, only 3 per cent of respondents identified investment as a motive, while a further 21 per cent mentioned saving. The proportion mentioning investment rose in the 1980s and for higher social classes, but even for the most affluent groups at the peak times hardly exceeded 12 per cent of responses. In contrast, half the sample gave traditional kinds of reasons for owning – its security, its appropriateness to marriage and childraising and the choice of housing and locations it offers. Even more numerous were reasons which identified (but did not explain) the superiority of owning to renting, or reactions to circumstances – low prices, availability of tax relief or right to buy subsidy and even 'it was the automatic choice'. Overall, the reasons given do not seem very compelling: households are clearly convinced that owning is better than renting, but are not very articulate in explaining why. They expect to accumulate capital but do not deliberately set out to do so.

To try to clarify what people are seeking from home ownership, households who had moved within owner occupation were asked what had persuaded them to move. Since second-time and subsequent purchasers are less concerned with overcoming constraints to access or with the change of tenure *per se*, it was hoped that people would be more conscious of what they hoped to achieve. Sure enough, only 9 per cent said they had reacted to circumstances, especially the need to retrench, and nearly half the respondents stressed that the new house was bigger, better or in a better area. About 15 per cent, said they moved for family reasons, including the breakdown of a relationship as well as new relationships or children. A distinct reason was to be nearer a job: this was given by 14 per cent of respondents and was often a reason for third or subsequent purchases. Only 4 per cent had moved to make financial gains, and more than half of those spoke in terms of increasing their investment rather than cashing it.

All the open-ended questions suggested that investment was not a major reason for house purchase. This was confirmed by questions about attitudes to finance. Extraordinarily, half the sample said finance was not a major factor. A further 20 per cent said they wanted to buy a property which would hold its value and protect their investment. Only 4 per cent said finance was a major factor. Of those who identified specific financial motives, many more said they wanted the cheapest property or to limit costs than wanted to maximize capital accumulation. Owners who had had three or more homes were more likely to say that finance was unimportant, less likely to opt for the cheapest property, but only 3 per cent said they wanted to maximize capital accumulation.

The survey evidence consistently shows that people become homeowners from a combination of specific advantages with general belief that it is the superior tenure. The pro-ownership rhetoric of the 1980s may have contributed to that belief, but its stress on the investment benefits of housing had relatively little effect on households. Indeed, many more people were concerned to reduce costs or protect their investment than to invest more: attitudes to finance seem more defensive than acquisitive. This finding cannot be explained away by the timing of the survey. Even in the depths of recession, a large majority continued to believe that in the long term house prices would rise. However, an even larger majority disagreed with the idea that one should buy the most expensive house possible in order to maximize capital gains. Overall, both answers to questions and actual behaviour make it clear that people are aware that home purchase carries risks as well as benefits and that they try to act in ways which diminish that risk. However, this sense of caution did not prevent many households from buying near the peak of the boom and many hundreds of thousands have suffered financial loss and even homelessness as a result. From them, later buyers have learnt that the risks are far greater than anyone realised in the 1980s. That has been the main cause of the slump in the 1990s.

The housing market of the 1990s had very much more complex relationships to identity and disadvantage than was the case 20 years before. Instead of a minority tenure conferring predictable benefits to both consumption and asset gain to married couples of middle to high income, owner occupation is now a very heterogeneous tenure, ranging from former renters of council houses and low-income Asian households, through a variety of single- and multiple-earner households to the very rich. For some, owner occupation has meant foregoing luxuries and holidays, postponing parenthood or dramatic financial reverse. For others, it has given a boost to asset-accumulation and in some years brought greater gains than savings could ever yield. The contrasts with non-owners also changed: no longer does council renting offer decent housing at low cost for skilled workers. Here again, housing exerts a multiplier effect on change elsewhere: unemployment or other misfortune can lead to renting substandard accommodation, often at comparatively high cost to individual or state, or to homelessness. Subsidy to right to buy gives households in continued employment an asset boost, but failure to keep up payments leads to temptation or requirement to sell and a struggle to find alternative accommodation. Perhaps most unpredictably of all, the housing market collapse has created a new privately rented sector as owners unable to sell are obliged to rent their home when they move elsewhere, and perhaps to become renters themselves in their new area. This sector is probably only temporary, as a revival of sales volumes and prices will prompt sales of currently rented stock, but the revival of private renting epitomizes the new uncertainty of the housing market. Whereas, in the past, housing was largely an outcome of other identity-forming experiences, it has now gained a larger, though very unpredictable, degree of autonomy over identity, consumption and asset gain or loss.

A new mix of regional uncertainties

In so far as there has been no simple, unambiguous shift to a new set of middle-class identities around enterprise and competitive individualism, so too there has there been no complete set of changes around jobs and housing which has led to a more precarious existence across the south east. None the less, there is a very real sense in which many different social groups in the region now consider their economic and social lives to be riskier than before the 1980s' 'boom and bust' scenario. The insecurity of identity, the new mix of uncertainties brought about by the neo-liberal form of 1980s' growth has, for different groups, fractured and made provisional their idea of a career, together with the assumption that housing 'stages' would track the family life-cycle.

Much of this material and symbolic insecurity is caught in Ulrich Beck's account of a *Risk Society* (1992), where risk is something that is said to permeate all aspects of our lives, from how we place ourselves to the institutional settings of work, family, class, knowledge and the environment at large. On this rather general view, we are perceived to be in the process of moving from an industrial society organized around the pursuit of wealth and goods to one organized, globally, around hazards and risk. Central to this argument is that it is the process of modernization itself, in particular the reflexive practices and forms of institutional rationalization, which are dissolving industrial society and paving the way for a 'risk society'. Leaving to one side the immanent logic of this view (see Rustin 1994), there is none the less something about Beck's ideas which resonates with the regional mix of uncertainties described, and the impact of neo-liberalism in particular.

Turner (1994), for instance, has made a strong case for locating Beck's account of risk in the context of the decline of social Keynesianism and the impact of free-market strategies more generally. In particular, he draws attention to two aspects of Beck's conceptualization of risk which chime with neo-liberalism. The first aspect concerns the unforeseeable nature of risks thrown up by the institutional practices of governments and markets which can no longer be adequately controlled by existing political regulatory measures. Thus, in the context of a neo-liberal heartland, the increasingly precarious nature of work and housing 'careers' – where the ability to plan long-term can no longer be automatically assumed even among middle-class households – may be traced to the specific political strategies of neo-liberalism and their unanticipated effect upon the workings of both labour and housing markets across the south east. To take but one of the more obvious examples, deregulated labour markets, unpredictable working lives and increased homeownership do not mesh. Their effects are incompatible and serve to disrupt assumed elements of middle-class culture and identity in particular.

A second aspect of Beck's argument highlighted by Turner is perhaps best understood as an extension of the first in that risk is defined in the broadest of social senses, rather than in narrow legal or entrepreneurial terms. Risk and uncertainty in social life are related to customs, practices, traditions and habits which act as some form of social regulation to filter out anxiety and to frame expectations. Thus, as certain assumptions about a 'job for life' or a 'planned career' or that housing is an investment linked to 'stages' in the family life-cycle lose their hold upon social expectations, the outcome for different groups is a mixture of opportunities and anxiety depending upon the skills and resources at their disposal.

For those at the bottom end of service labour markets in the growth spots of the 1980s' south east, we have already seen how exposure to a more risk-laden (read 'flexible') labour market can affect both work-status and identity, as well as compound material insecurity (see Allen and Henry, 1996, 1997), but even those directly 'included' by the neo-liberal project were not immune from greater social risk. One of the sharpest illustrations of this is to be found in the unforeseen workings of the owner occupied housing market across the region in the 1980s, and the promotion of homeownership as a sign of privatism and identity.

The passing of a set of middle-class assumptions and values attached to the security of home ownership is none the less only one amongst a new mix of uncertainties brought about by a particular form of growth and its geography. It is all too easy to overstate their effects, or indeed their social institutionalization, as Beck does, but there is a sense in which a culture of risk and uncertainty became more embedded in the social fabric of the south east than elsewhere in the 1980s. Indeed, the economic risk and enterprise inscribed in neo-liberalism may have exacerbated the regional mix of social and economic uncertainties. Above all, the instability of the project and its particular geography is there to see with the benefit of hindsight in the 1990s, and perhaps its self-defeating character.

Part III

SPACE–TIMES OF NEO-LIBERALISM

5

SELF-DEFEATING GROWTH?

Throughout this book we have reflected on the experience of the south east of England in the 1980s and 1990s. However, we have not simply been concerned with that history. Instead, we have drawn out some key issues about the ways in which regions are defined and define themselves. We have stressed the importance of cross-cutting relations between regions, both within the countries of which they are a part and across their borders in wider networks of regions and places. At the same time we have explored the significance of relationships between places within and beyond the boundaries of particular regions. One of the implications of this is that regions are not pre-given entities handed down by some immortal map-maker from on high (or even from the computer systems of the OPCS), but are the products of social processes of interaction and representation. Their definitions are likely to change over time.

Equally important, and in this respect the south east is a particularly powerful case, regions are the products of overlapping social relations which stretch across space with little respect for the boundaries we give them. Instead of viewing the south east as just another patch in the quilt which makes up the UK's (or Europe's) regional system, it is important to acknowledge the importance of relationships and linkages which stretch beyond and cross-cut its, already rather porous, boundaries. The region is defined by networks of relationships and interconnections which stretch across the UK, Europe and the wider world. It has no independent existence separate from them.

As well as exploring the ways in which regions are made and remade, defined and redefined across time and space, all of the chapters of this book have explicitly or implicitly questioned the notion of growth which has come to dominate mainstream political debate over the last two decades, whether associated with moves towards what Jessop (1994) has called a 'Schumpeterian workfare state' or the emergence of a new political settlement which Hay (1995) has labelled 'Blaijorism'. On all sides of this debate it has generally been taken for granted that 'growth' will bring all of the other things we want in its wake – jobs, income, wealth, even welfare provision and the resources to pay for environmental programmes. Growth, according to this approach, is quite simply a 'good thing'. Even the concentration of growth in particular places is often presented as a (not always unwelcome) necessity because eventually it can be expected to feed through the whole of the economic system. Ultimately, it is argued – particularly in the neo-liberal or 'Thatcherite' version of history – that everybody who deserves to will benefit, even if they may not all do so to quite the same extent or at the same time.

Such an approach is fundamentally flawed because it fails to consider the nature of economic growth, its differential impact in particular places and regions and the people who live in them. It fails to take account of what might be called the political and economic geographies of growth. A focus on the south east of England in the 1980s and first half of the 1990s highlights the need to explore the tensions, inequalities and contradictions which exist within specific forms of growth. Even within the 'growth region' of the south east, as the last two chapters have indicated, there have been quite distinctive and different responses to growth (or in some cases decline) from different social groups and in different places. The 'growth' of the south east has helped to generate new lines of inequality both within the region and between the region and other parts of the UK. The predicted sharing, or 'trickle down', of growth has not taken place, but – and in the context of the 'two nation' strategy identified by Jessop *et al.* (1990) possibly more important – the dominant form of growth has been self-defeating, in the sense that it has been undermined by its own apparent success. Instead of providing a generalizable model capable of being spread across the UK, the growth of the south east in the 1980s was based on decline and stagnation elsewhere. The definition of the south east as a growth region in the 1980s was predicated on fundamental regional disparities within the UK (and even within the south east itself), and it was precisely this unevenness which made it impossible to sustain regional growth beyond the end of that decade.

Unequal and fragile growth

The growth of the south east has proved to be both unequal and fragile (CERN, 1993). It has reinforced inequality between regions, as well as being experienced unevenly within the region. The emergent inequalities which helped to define the south east's growth also served to undermine it in the 1990s, not only enfeebling the regional economy but also bringing the possibility of national economic growth to a stuttering halt (see, for example, Peck and Tickell, 1992). As we have seen, one aspect of this inequality was celebrated politically in the 1980s and early 1990s, since it seemed to confirm the necessary 'superiority' of Thatcherite growth (based on finance and services) and the 'backwardness' of the old primary producing, manufacturing and Labour voting areas outside the region. But embedded within these divisions were the mechanisms of fragility which helped to undermine the south east's position in the 1990s. The very propulsive growth dynamics that were identified in Chapter 1 incorporated within them the factors which in turn undermined the growth they generated.

Despite the political rhetoric of the 1980s the south east could never have been a model for the rest of the UK: on the contrary, the relative decline of other regional economies was a necessary corollary to its growth. It was a 'core' with rather peculiar linkages to the rest of the British social organism. One key aspect of this was the extent to which, through the late 1970s and 1980s, in effect the political economy of the south east as 'growth' region operated in ways which restricted the possibility of growth in other regions of the UK. The south east was more than just a spatial expression or consequence of Thatcherism's 'two nation' politics, not least because its formation was fundamental in shaping the politics of neo-liberalism in the UK. These politics were always spatially constituted.

Massey (1988b) charts the ways in which the uneven development of the UK economy has been self-reinforcing and cumulative since the mid-1970s. Not only was there a growth of employment and economic activity in the new sectors such as banking and financial services, but the south east's share of activity in manufacturing industry also increased both in terms of output and employment. Of course, as Massey argues strongly, this was not the result of some necessary and inexorable logic, but the outcome of social and economic processes which need to be questioned and explored. Aspects of economic change which shape what is possible are too often simply taken for granted. For example, Fielding (1992) notes the extent to which during the 1980s, the south east operated as a drain on the market for highly qualified labour, leaving other regions with less of the highly skilled labour on which growth might be based. In other words, the south east was a 'growth' region at the expense of the other regions of Britain.

The concentration of particular forms of growth in the region – at the expense of other regions of the UK – had the effect of substantially increasing labour and other costs, generating skills shortages and making it more difficult to resolve them. The operations of the housing market provide only the most obvious example of the negative consequences of unequal growth. Because growth was concentrated in the south east, housing costs rose dramatically there. This made it difficult to attract skilled labour from elsewhere in the UK, since the gap between housing costs was not covered, even by the significant wage differentials between the south east and other regions. Paradoxically, the south east succeeded (as Fielding notes) in attracting skilled labour (particularly highly educated labour) away from other regions (so that their economic bases were weakened), while at the same time labour costs in the region were forced to levels which were unsustainable and it became difficult to attract all the labour which was required.

As has been indicated in earlier chapters, the concentration of 'growth' in the south east was accompanied by a continued shift towards economic activity and employment in financial and business services, and a shift away from manufacturing. The rising real incomes of most of those living in the south east were accompanied by a dramatic increase in spending based on consumer credit, itself fuelled by a rising housing market in the region (see Hamnett and Seavers, 1994a). Consumer demand in the region rose significantly and (since the manufacturing base of the UK had been substantially eroded) encouraged the large scale import of manufactured goods. The inevitable consequence was the growth of balance of payments deficits at the start of the 1990s, which in turn heralded the collapse of the UK's short-lived boom.

In the 1980s a rising housing market gave existing residents in the south east the promise of increasing capital gains, but the collapse of the market at the end of the decade brought sharp reversals even for many of those residents (particularly recent entrants). Outside the region few people were seriously affected by 'negative equity', yet it became a major issue on the national (as well as regional) political agenda in the early 1990s. This was a reflection of the political significance of the south east as discursive base of a neo-liberal project for the UK as a whole, which promised continuing (and almost effortless) prosperity to those who bought into it. An inability to deliver the goods in the Thatcherite promised land, therefore, dramatically called the project as a whole into question.

Not only did the promised gains prove illusory, the basis on which they were constructed (i.e., through the emphasis of sharp differentials with other parts of the country) also helped to generate the wider economic reversal. The south east's 'success' helped to price it out of the market, through the 'overheating' of the regional economy. The example of housing is only one expression (or symptom) of the fragility of the south east's growth as a consequence of uneven and unequal development within the UK. The costs of commercial and office property in London, particularly the City, reflect its separation from the wider UK economy, since they are much more linked into the requirements of a global financial system than those of industry in the UK (see Hutton, 1995). But this means both that other forms of investment are effectively discouraged and that if the financial sector faces problems (as it did in the late 1980s) then its localized impact is significantly greater. The economic base turns out to be rather narrow, precisely because of the extent of concentration in a relatively limited space, with the consequences feeding out to the suburban heartlands, whose residents depend on commuting to London for their incomes.

The concentration of other service activities in the region (including airports and ports) inevitably forces up costs and introduces rigidities, as well as encouraging labour shortages in areas of particular skills. The structure of the south east as it was (re)constructed through the 1980s began to act as a self-limiting bottleneck, not only on its own growth, but also on that of the UK as a whole. For a time, this encouraged feverish activity within the region, confirming its pre-eminent position and bringing increased income to its residents, but it also brought in its wake the danger that the region will lose its international competitive position. This was a particular problem for the south east since – as Murray (1989) has argued – its main selling proposition externally is that it offers a deregulated, and hence cheaper, source of labour and other facilities, particularly in financial and other services. One consequence of unequal growth in the UK being focused on the south east in the 1980s has been to undermine its international position by increasing its labour and infrastructural costs.

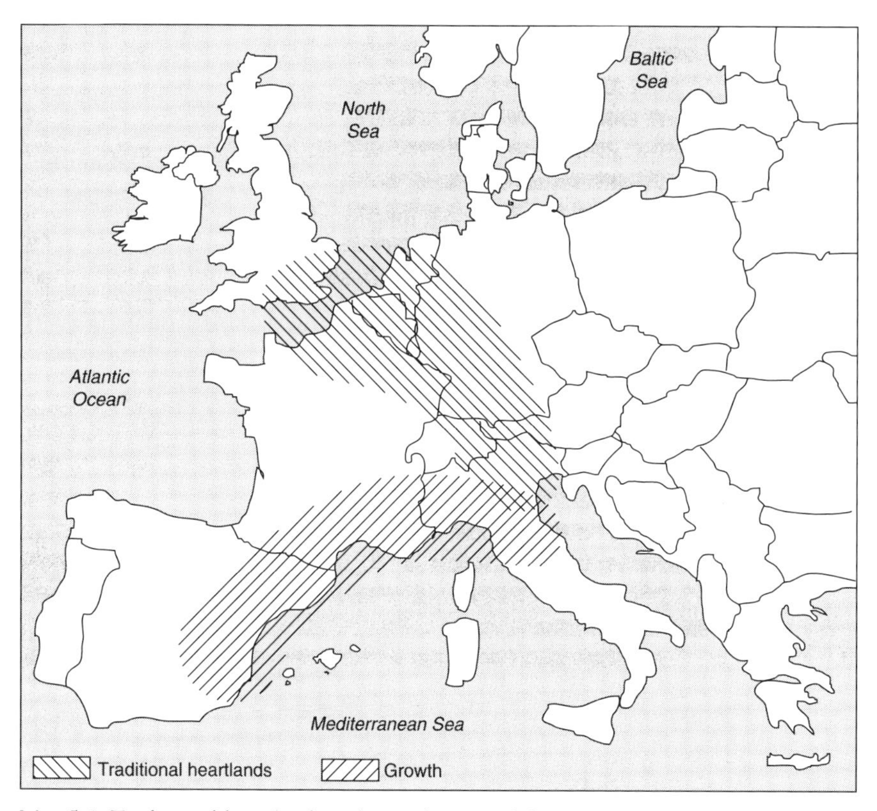

Map 5.1 Traditional heartlands and growth areas of the European Community
Source: Commission of European Communities (1991)

Competing on cost leaves the south east in an increasingly weak position in the face of other European regions with still cheaper cost bases.

Despite its dominance in Britain, the position of the south east in a wider context is by no means so secure. Within the UK it may still be the most prosperous region, but its position within Europe is much more ambivalent, particularly since its main selling point is as a relatively cheap and deregulated offshore financial haven. Even if the City is likely to remain secure at least in the medium term, because of its role within a set of global financial markets, the rest of the south east is in danger of being marginalized on the edge of a much larger western European economy. The strategy for 'Europe 2000' (Directorate-General for Regional Policy, 1991) sets out to provide a European context for spatial development planning at national level and below. It predicts that there will be two main areas of economic growth in Europe, with a new crescent (or 'banana') of growth stretching around the Mediterranean from northern Italy halfway down the coast of Spain linking to what the report calls the traditional core which stretches from northern Italy to the south east of England. Such a pattern is – of course – worrying for those areas outside the golden heartlands (including most of England and all of Scotland, as well as central and eastern Europe) and the report highlights the likely unevenness of development between different parts of Europe, moving beyond the existing boundaries of the European Community to show what the consequences are likely to be in other countries, too. The south east of England is only at the extreme tip of the banana, apparently in danger of marginalization if growth begins to concentrate closer to the new crescent, or if expansion to the east generates a new corridor of growth (with German unification, the entry of Sweden, Finland and Austria to the EU, and the opening up of central Europe and Russia to the market). This carries with it a clear threat to England's south eastern place in the economic sun, which may help to explain the growing political ambivalence of Conservative politicians representing constituencies in the region to the institutions of the EU.

The internal fragility of a 'growth' region

If one aspect of the UK's economic growth in the 1980s was a reshaping of its space economy, albeit one which superficially simply seemed to confirm the continued dominance of the south east, then an equally important reshaping took place within the region. This also contributed to the fragility of regional growth. The metaphor of the doily which is developed in Chapter 2 is fundamental to an understanding of this process. The region is defined by the holes within the general image of growth, as much as by the relationships and connections between places and across space which take the imagery of growth for granted. Some of the emergent divisions are clear enough and well known (see SEEDS, 1987). The contrast between London's inner cities and its leafy suburbs or gentrified areas is an example whose force is only confirmed by a consideration of the greenbelt enclaves of the outer south east, such as Buckinghamshire – as outlined in Chapter 3. Divisions between the old Fordist (and even pre-Fordist) manufacturing centres, such as Rochester, Slough and Luton, and the new (high-tech) manufacturing centres such as Cambridge and Swindon, are easily apparent.

But some equally important divisions within the region are rather more easily missed, as is the complexity of some of the divisions which are frequently noted. In particular, for example, there is a marked contrast between those places which are largely dependent on City-based financial services and which experience growth through commuting, and those in which growth is actually taking place. South of London – the south east of the south east – is increasingly made up of dormitory towns and commuter villages with little locally based employment. Indeed, the traditional industries of the area are in decline whether in the primary (mining), service (tourism) or manufacturing sectors. The Medway Towns have an enterprise zone and the east Thames corridor (now relabelled the Thames gateway) has been identified as a focus for major urban regeneration schemes, while the tourist resorts of Thanet and their associated ferry services appear to be facing inexorable decay (to the extent that east Kent has been identified by the EU as an area deserving of financial support), but the rest of Kent is largely based around commuting to London, with the most intensive recent political debate centring around resistance to successive proposals for rail routes connecting London to the Channel Tunnel.

The region's growth dynamics help to reinforce broad divisions which are only invisible because they are taken for granted. The growth dynamics of the 1980s tended to bypass the old inner cities and the region's older manufacturing towns and cities, instead encouraging growth pressures in places where they were not always welcome, and in places (such as Cambridge and even Milton Keynes) where labour market and other limits helped to reinforce the downturn when it came. The inequality and unevenness within the region which underlay growth helped to ensure the boom's fragility, although in ways which ensured that the recession of the early 1990s hit the south east harder than other parts of the UK, with a decline in employment across the board.

Little attempt had been made to train or retrain the pools of unemployed/ low-skilled or 'wrongly' skilled people to fill the demands for skilled labour of particular types, or to reuse the places which had been left high and dry (CERN, 1993). Places, such as Docklands, which received massive investment at the expense of other areas of London (and incidentally the rest of the UK) turned out to be dependent on forms of growth which did not survive the recession. Indeed, to have taken any of these issues seriously during the 'growth' years would have been to question the very discourse of a growth region on which the south east's social and political dominance relied. Within that discourse, the south east was a 'growth' region precisely because of the apparent absence of state intervention or 'interference', regulation or investment.

One result of this is that today some places (such as Luton, but also Brighton, Thanet, and the Medway Towns) which are part of the south east region, if only as holes in the doily, have levels of unemployment which are significantly above the national average, but even those which rode the crest of the wave during the boom years were hit hard when the slump came, with, for example, unemployment in Milton Keynes in the early 1990s rising at a rate still faster than that of the south east as a whole where it was rising faster than in the rest of the UK. The point here, of course, is not to shed a few crocodile tears about the problems of the south east, but simply to indicate some of the ways in which it was (socially and discursively as well as materially) constructed in the 1980s and the effects of those processes on the recession when it came.

The Lawson boom of the late 1980s, as noted in Chapter 1, helped to define the south east as a 'growth' region, and was itself largely defined in popular imagination by imagery drawn from the south east, whether in terms of 'Loadsamoney' wide boys, in the 'greed' of City speculators in red braces and striped shirts, or in the quasi-rural mythology of Home Counties suburbia (see Thrift, 1989). But this, as Chapters 3 and 4 indicated, too, helped to highlight its fragility. The global collapse of stock markets and losses in insurance markets in the late 1980s had particularly sharp impacts in London because deregulation left more at risk. Mountains of debt could no longer be serviced either by major development companies or by domestic householders. Interest rates rose, the property market collapsed and workers in the financial sector were laid off or saw their incomes fall dramatically (Hutton, 1995).

Instead of housing representing a safe investment in 'bricks and mortar' for many people (see Hamnett and Seavers, 1995), living in the south east began to imply an increasingly threatening financial drain. As noted in the previous chapter, the fear of 'negative equity' swept through the previously secure suburban bastions of middle-class respectability, although it hit high-profile symbols of property-led renewal (such as London's Docklands) first. Those who had bought property in the 1980s and earlier were affected particularly dramatically, as the downside of the deregulation of housing finance hit hard. Building societies began to repossess the properties of those who could no longer afford to pay mortgages agreed on the basis of optimistic assumptions about household income, employment and interest rates. The flood of repossessed houses onto the market also helped to force prices down, so that the monthly payments of existing mortgage holders were often rising at the same time as the value of their property was declining.

The power of deregulation as an ideology of 'growth' in the 1980s also brought with it an opposition to infrastructural investment through the public sector. In practice, this still allowed the south east to call on the biggest share of the infrastructural investment which was taking place (for example, in the form of the M25 and the large sums poured into London Docklands), not least because there was a clear move away from any regional policy which might have been expected to favour Britain's older industrial regions (see Massey, 1988b). But even this investment was hopelessly misdirected. Transport infrastructure was inadequate to meet the demands either for commuting or freight (see Church *et al.*, 1992). Investment in roads increased traffic and led to endless jokes about the M25 as the 'largest parking lot in Europe', while the Docklands experiment proved unable to shift the City eastwards (however successful it was in moving Fleet Street to Docklands) and had little positive impact on the lives of those living in the Docklands boroughs. Most obviously, the commitment to deregulation left Britain in a weak position to take advantage of the Channel Tunnel, retaining an inadequate set of rail linkages (in sharp contrast to the French experience) and only gradually moving towards some wider set of initiatives (for example, through proposals for the east Thames corridor/Thames gateway). From being the main beneficiary of the deregulationist and privatization strategies of the 1980s, by the end of the decade, it was clear that the south east's link into those strategies brought its own disaster in its wake.

The changes of the 1980s were characterized by the reinforcement of inequality through processes of uneven development both within the south east and in its relations with other parts of Britain. The divisions widened markedly across the decade, only narrowing with recession and a greater equalisation of economic misery. At the same time, some features of unequal growth were among those which helped to tip the regional and national economies into recession. These included highly localized labour shortages, differential housing costs, high levels of personal debt, failures to meet basic infrastructural requirements and the dominance of employment based around activities which relied on the continued growth of personal debt of one sort or another. If the south east's growth was the product of government inspired deregulation across a range of sectors, then the form taken by its recession also reflected the consequences of that deregulation.

The state has always played and continues to play a key role in shaping patterns of growth in the south east, which – as we have seen – was a privileged beneficiary (indeed almost by-product) of government policy in the 1980s. There is nothing 'natural' or 'necessary' about the shape taken by that growth. The south east is not even (and never has been) simply the outcome of market processes, or of deregulation, despite the dominant rhetoric. The form of growth which took place in the south east was fundamentally influenced by state intervention and emergent forms of regulation, even if they sometimes appeared in the guise of 'deregulation'. State policy has been fundamental to the construction of the south east as 'growth region'. The paradox is, however, that there is little or no evidence that an active or explicit regional policy was being pursued either by national government or by other state agencies. Although the implication of dismantling existing forms of regional policy at the start of the 1980s may have been to 'benefit' the south east, that was never stated to be official policy. Nor was 'deregulation' coupled with infrastructural investment ever presented as a regional policy. There was no significant regional infrastructure within the region. Only SERPLAN (the south east's regional planning forum, made up of representatives of planning authorities) has had any role, and its main purpose seems to have been to provide an arena within which planning authorities are able to negotiate about pushing growth away from their backyards. For a time, the only place in the region identified for significant growth was Milton Keynes, although the slump brought with it a new interest in regeneration along the north banks of the Thames in Essex and Kent and into London's east end. The only unified agency which might have been expected to have a significant influence – the Greater London Council – was abolished in the mid-1980s.

The south east exhibited few of the features of what Amin and Thrift (1994) have called 'institutional thickness'. By this, they mean a combination of a strong local institutional presence coupled with high levels of interaction between those institutions, alongside the possibility of coalition-building or domination by some groups which makes it possible to move beyond sectional interests (therefore socializing costs and overcoming 'rogue' behaviour). Amin and Thrift highlight the importance of the development of a 'mutual awareness that' the institutions 'are involved in a common enterprise' (ibid.: 14). In this context, institutions are taken to include a wide range of agencies, from firms and chambers of commerce to local authorities, civic bodies and trade associations. The networks of interaction will tend to encourage the development of relations of trust and, when working positively, are likely 'to stimulate entrepreneurship' (ibid.: 15) (for related arguments which build on the notion of social capital, see Putnam, 1993).

This perspective fits uneasily with the agenda of neo-liberalism which underpinned the notion of south east as 'growth' region. Even in those places within the region where 'institutional thickness' might have been significant, the growth was predicated on questioning and undermining it. So, for example, it was the questioning of old assumptions which fuelled the financial boom of the 1980s, the breaking up of the old boy networks of trust (even if the same people often benefited financially), the entry of new players into previously protected marketplaces. Many of the 'hot spots' – like Milton Keynes, Bracknell and Swindon – were weakly developed in institutional terms, while in others, such as Cambridge, the relationship between the existing institutions and the new were always much more attenuated than the 'hype' suggested. As a result, not surprisingly, Amin and Thrift directly pose the question: how 'do we account for the success of areas like the south east of England . . . where institutional "thickness" . . . has not been readily apparent?' (Amin and Thrift, 1994: 17).

If the earlier chapters of this book have helped to answer that question by identifying those elements which have made the south east a 'growth region', the corollary in this chapter is to confirm that the lack of institutional thickness (at regional level) has helped to undermine the possibility of stable growth. Institutional thickness may not be a requirement for growth, but its absence certainly appears to make it difficult to sustain. A fundamental aspect of the south east's failure to sustain growth, either within the region or as a driver for the UK as a whole, is to be found in the limitations of its institutional base – its lack of a developed set of interlocking networks. In other words, its failure is the failure of a neo-liberal approach which believes that no linkages are good linkages. Peck and Tickell make a similar point in highlighting what they see as the absence of an effective local mode of social regulation. They note that, 'The region enjoyed strong economic growth as a result of its privileged place in *national* regulation strategies and *international* accumulation strategies, but subsequently was shown to lack appropriate *regional* regulatory mechanisms for the sustenance of growth' (Peck and Tickell, 1992: 359; emphasis in original). Although we are less convinced than Peck and Tickell about the possibility of identifying specific 'local modes of social regulation', they are persuasive in highlighting the importance of regional weaknesses in explaining the south east's slide into slump.

Looking for a regional policy

It is sometimes easy to forget that regional policy in the UK has not always meant encouraging inequality, whether to provide incentives to others or to provide models for the less successful to emulate. In the 1960s and 1970s, building on a longer tradition going back to legislation of the 1940s, stress was placed on the dangers of 'overheating' and the need to decentralize activity away from the south east towards regions in which there were pools of unemployed labour. The political strategies of the Wilson era were based on a similar commitment as that of Thatcherism to the success of the UK operating within a capitalist world market. In the earlier period, however, emphasis was placed on the UK as a corporate entity, with the state acting as a sort of management team. Since the 'overheating' of the national economy was perceived to be a problem, more equal growth was seen as a potential solution. One result of this was that in the late 1970s the gap between the south east (and the west midlands) and other regions of the UK was narrowed – at least partly in response to an active regional policy (Moore, Rhodes and Tyler, 1986).

The new government elected in 1979 argued, by contrast, that regional policy had operated as a dangerous brake on initiative in the more active and dynamic areas of the economy. The dangers of 'overheating' seemed to be exaggerated because the economy as a whole was in serious trouble. Not only were there substantial pockets of unemployment in the core areas, particularly in the inner cities, but some previously prosperous regions, such as the west midlands, themselves seemed to be facing decline. The answer, therefore, officially at least, was to move towards more explicit targeting and to release the entrepreneurs, and the City, from the controls which limited their options. As we have seen, that strategy, too, has run into the ground: even within the heartlands of the dream in the south east.

It is, therefore, important to think again about ways of challenging and questioning the political and economic 'common sense' which manages to generate inequality without bringing the more widespread growth which is supposed to go along with it. Implicit and, sometimes even, explicit challenges to the neo-liberal 'common sense' which dominates in the UK have begun to emerge from policy debates around Europe and local government.

Regional policy may have been removed from the respectable language of British politics, but it remains fundamental to the strategies of the European Commission. A programme of structural readjustment is central to the project of building a single European market based around economic convergence between the regions of the European Union. The EU's regional policy (organized through the European Social and Regional Development Funds) is an explicitly economically liberal one in the sense that it is not predicated on an attempt to achieve equalization between regions for its own sake. On the contrary, its underlying principles are concerned with the removal of obstacles to the free flow of labour and capital as well as goods within the EU (see Cutler *et al.*, 1989). The sharply different levels of economic development between regions of the Union make it more difficult to achieve the aim of economic integration, because they encourage the collection of pools of low-paid and unskilled labour in countries and regions dominated by agricultural employment (such as Greece, Ireland and Spain). At the same time, one of the strategic priorities of the European Commission is to encourage the restructuring or 'reconversion' of those regions whose economies have been based on traditional primary and secondary industries. Investment in new infrastructure and in skills training is supported through the EU's regional funds in these areas, with the intention of fostering restructuring, that is reducing capacity in the old industries and fostering the growth of new ones (in high-tech sectors and in services).

In their analysis of the institutional issues facing regions in Europe, Amin and Thrift (1995) suggest that the main driver of the policies and policy framework developed by the European Union is essentially a form of neo-liberalism. Their description looks very like a form of Euro-Thatcherism, and as a result they emphasize the extent to which current policies are likely to generate and sustain substantial divisions between regions. They outline clearly the growing regional divide within Europe, which is masked by the language of cohesion and convergence to be found in the official documents underpinning the notion of a 'Europe of the Regions' (see also Dunford and Kafkalis, 1992). The impact of the single European market and the opening up of local and protected markets to competition helps to undermine and to reinforce the position of the stronger regions. European social policy is aimed at enabling the free movement of labour, rather than providing welfare support to individuals (see Cochrane, 1993a). If anything, it serves to protect the wealthier regions. Amin and Thrift argue that heightened competition is the main effect of EU policies, while the specifically regional policies offer little prospect of reversing them. In effect, we are seeing the reinforcing of a sharper regional hierarchy.

> Liberalisation, the rolling back of the welfare state, deregulation and the unification of markets on a European scale will serve only to transfer and consolidate resources – from capital and investment through to skills, knowledge and entrepreneurship – towards the core regions which offer the highest returns.
>
> (Amin and Thrift, 1995: 47)

The EU has also, however, been the focus of arguments for the revival of a more managed approach to regional policy, much closer to the assumptions of UK policy in the 1970s. If one model is Euro-Thatcherism, the second is close to a Euro-Keynesianism (or even what might be called Wilsonism) – encouraging moves towards equalization through expansion, through public expenditure and fiscal policies. Amin and Thrift summarize a number of aspects of this approach, including suggestions for 'federal fiscalism' which would involve the distribution of tax income to 'less favoured' regions to help with the financing of development. It therefore, moves beyond neo-Keynesianism because it also assumes the development of more targeted industrial policies designed to promote innovation, networking and best practice (see Murray, 1992). Despite some sympathies for this approach, Amin and Thrift believe that it is likely to be unsustainable, particularly because it fails to recognize the extent to which regions are 'the product of unique development trajectories, rather than the more or less imperfect reflections of any "ideal" growth model' (Amin and Thrift, 1995: 49). This conclusion fits well with our own analysis of the south east as 'growth region' and reinforces the need to explore more critically the strategies which are on offer.

Looking for a way forward

There are two main ways of approaching the challenges which face policy-makers in the field of uneven development on a global scale, particularly as it finds its expression regionally and locally. The first is the one that was adopted in Britain in the 1980s, namely, a commitment to deregulation and a belief not only that there is little that can be done in the face of global economic forces (the market) but also that – in the end – removing obstacles to the operation of the market will be for the best in any case, for everybody. We have argued against this approach, not only highlighting some of the negative consequences of deregulation but questioning the claims of its supporters that it will help to generate stable growth – even in favoured locations such as the south east. We have also, incidentally, highlighted some of the ways in which 'deregulation' as a government policy may actually mask forms of subsidy and support in the name of the market. The deregulationist approach to cross-national initiatives is one which seeks to expand and extend the market across boundaries, in ways which undermine forms of labour protection and guarantees – for example, in the form of the North American Free Trade Agreement (NAFTA) which has a minimal institutional expression since it has little regulatory policy to develop and protect.

The second approach is one which recognizes the importance of social regulation and economic interdependence as a basis on which to develop appropriate strategies in a global context. Accepting the importance of developing an effective regulatory regime, of course, can only be a starting-point. The current policies of the European Commission, for example, clearly stem from the recognition that some form of cross-national regulation is appropriate, largely on the basis that it is necessary to underpin the extension of a competitive market within Europe (to break down existing barriers) and also to help build a more competitive European economy within global markets. It is recognized at European level that leaving very large gaps between the economic standing of regions within the European Union may undermine some of the stronger economies, instead of strengthening them. But the emphasis of European regional policy is on infrastructural spending – allowing and encouraging development which leaves the poorer regions dependent on the richer ones (with a spread of branch plants, a growth of tourism and an encouragement of agriculture based monoculture). For all the rhetoric of 'social integration' and 'social inclusion', what is promised is a regulatory framework aimed at reproducing forms of uneven development.

Nevertheless, the changing structures of the EU are beginning to offer – or at least to point to – the possibility of a regionally based politics, which may be able to generate alternatives and to construct different visions of the future. Although it is important to recognize that in large part the policy direction taken by the EC reflects the institutional need of the organization to define a role for itself separately from the national governments which constitute the membership of the EU, it is also important to recognize that European initiatives reflect a changing world of regional arrangements and an attempt to get to grips with them. It is increasingly difficult to define regions in terms limited by the boundaries of nation states. The south east of England is a region of Europe as much as it is one of the UK: indeed, as Chapters 1 and 2 have indicated, its linkages with Europe and with a global economy are crucial in explaining its identity and role within the UK. In this respect, therefore, the creation of a Committee of the Regions within the EU's structures after 1992 also represents more than just a bureaucratic attempt by EC officials to bypass the power of national governments expressed in the Council of Ministers. It reflects a growing understanding that regional governments and political interests have to develop their own international networks and linkages if they are to play an active part in shaping their economic futures. It also reflects a growing understanding that initiatives taken at regional and local levels (within transnational networks) may have significant impacts – the view that regional economies are simply passive outcomes of the operation of global processes is one which is increasingly challenged by the experience of European regions which have developed active industrial and 'foreign' policies (such as Baden-Württemburg, see Morgan, 1994) In sharp contrast to the 'top-down' Wilsonian regional policies which were centralized and allowed neither for any significant local or regional flexibility nor for any significant democratic accountability, in principle, at least, the opening up of regional politics might encourage a wide ranging debate about the direction of change.

The appropriate political structures barely exist in the UK at present through which the necessary debates might be undertaken or the appropriate initiatives be implemented. If a serious alternative to existing policy is to be developed, it will be important to move towards the creation of democratically elected regional authorities with delegated powers to discuss strategic issues relating to economic development as well as transport and development planning and infrastructural development (other possible responsibilities might include policing, fire and health services). As the earlier chapters have indicated, of course, determining the political boundaries of such a region in the south east is not likely to be easy, not least because it implies the possibility of reflecting changing sets of social relations through the imposition of political boundaries which are likely to be frozen over time. But the difficulty of drawing precise boundaries should not represent a major obstacle: the allocation of disputed areas could be determined by local referendums, as is the case in Germany. Some of Germany's regions, such as Bavaria, take some of their legitimacy from history, but others, including the biggest, North-Rhine Westphalia, and the one with the biggest reputation for its economic initiatives, Baden-Württemburg, are the products of the post-1945 constitution. At present some three-way division might make sense (with a revived Greater London, and the ROSE divided along south east/north west lines), although we would not wish to propose any particular carve-up of territory. The existence of regional authorities of this type would not only help to create forums in which discussion about strategic issues could take place, but also allow those discussions to be linked to practice. Such authorities could negotiate with other major players in the region and outside, for example, in Europe, but also elsewhere in UK, helping to develop strategies which stretch beyond particular regions.

The existence of such authorities should not undermine a recognition of the importance of more locally based political initiative. To some extent, the potential of such activity can already be seen from the experience of the 1980s, since there have been no regional authorities, but local authorities have played a very active part in seeking to develop new initiatives, even in the face of opposition from national government. Many councils have set up their own European offices and almost all are engaged in developing some form of economic policy, whether to encourage or restrict growth. It is from local authorities that much of the initiative for economic development has come. At the same time a whole network of new agencies (including Training Enterprise Councils, Local Enterprise Companies and a wide range of public–private partnerships) has emerged, often with activist orientations, but also frequently with little interest in democratic accountability (see Cochrane, 1993b). A more formal and explicit recognition of the economic policy-making roles of councils, with adequate powers and access to resources, would make them capable of bargaining effectively with other interests over economic and other forms of development, particularly as part of a wider system linking them to regional, national and European initiatives, planning mechanisms and discussion forums. One task of local authorities on this model would be to open up debates more widely, allowing for the development of forms of participation which went beyond simple models of electoral accountability (and even narrow electoral boundaries).

Some local authorities within the south east have sought directly to move beyond the limitations imposed by central government, setting out themselves to reshape regional identities, and undermine emergent patterns of unevenness. Kent County Council, for example, has developed extensive linkages with Europe, even going so far as to join with regional authorities in France and Belgium to build a transnational Euro-region (see Cochrane, 1994). It is not necessary to believe that this 'region' has much more than a 'paper' existence to acknowledge that moving beyond national boundaries may provide a powerful means for local and regional initiatives to look for new ways of surviving in a highly competitive world. Such initiatives may also contribute to the ways in which regions may be able to define and redefine themselves in this context. If a first stage of local economic policy may be to differentiate at local level, place by place, as policies develop, the recognition of regional and wider linkages may become more important. Despite its institutional and bureaucratic rigidities (and its political weaknesses) the European Union has become a focus for the attention of a wide range of economically activist local and regional authorities from the UK and other member states. There has been a mushrooming of local authority offices in Brussels, and, more important, increasingly tight formal and informal networks of organizations (such as RETI – which brings together authorities from the declining industrial regions of Europe) and agencies have begun to emerge to take advantage of the limited institutional space which has been created (see Benington (1994) for a discussion of the emergent policy arena).

But it is not just the local governments with their appointed officials and elected politicians who are beginning to recognize some of the weaknesses of deregulation as a model for effective development and growth. Lock (1989) has argued forcibly for a form of (unelected) regional corporatism on which to base development planning, largely because he believes that the localised fragmentation of existing arrangements makes it difficult to unlock potential development land for large scale housebuilders. At the same time, despite their reluctance to embrace notions of regional government, even the Major government moved towards some acceptance of the need to have some sort of regional focus for its activities. Regional offices, bringing together officials of the main departments of government, have now been set up for all of England's regions, and – of course – Britain's sub-state nations already have their own institutional structures. In some respects, such arrangements may point away both from the traditional centralism of the British political system and from simplistic forms of deregulation, but they do so only by implication and only to a limited extent. It might even be argued that the formation of these offices represents another attempt to remove power from local government, by increasing the possibilities of regional surveillance and monitoring, and – of course – their officials remain formally committed to the pursuit of deregulation.

Moving beyond the 'growth' region

The perspective developed by Amin and Thrift (1994, 1995) is particularly helpful in this context because it highlights the need to move beyond a simple focus on the formal institutions of government, and the implementation of policies from above. Amin and Thrift set out to explore the policy implications of the notion of institutional thickness – recognizing the economic importance of a wider set of institutional networks – through an engagement with debates about democratic pluralist theories of associational politics (see also Hirst, 1994; Cohen and Rogers, 1995). The politics of associationalism build on the notion that society is made up of a plurality of associations, through which individuals come together in a complex variety of ways reflecting their multiple identities (at work, in leisure, at home, as consumer, and so on). These associations are not necessarily locally or regionally based. However, many of them will be and the ways in which they come together help to define regions in terms which recognize the importance of diversity rather than homogeneity. They offer the prospect of developing 'intermediate forms of governance', beyond the state. Building on the traditions of guild socialism, for example, Hirst (1994) is particularly interested in the ways in which associations might operate to provide a significant alternative to the centralizing and bureaucratic forms taken by the welfare state.

In this context, however, the key points are to identify (and encourage the development of) 'networks of intermediate institutions in between market and state' (Amin and Thrift, 1995: 50). The interweaving of relationships through networks should help to encourage joint working, the transfer of innovation and best practice. Crucially, however, in this model the process of institution-building, should be enabling rather than created by imposition from above. The emphasis is on organic change. There is no (even implicit) assumption that there is some norm to which each region should aspire. The fostering of intermediate forms should help to generate 'a process of collective governance of the socioeconomy' and should also make it possible to broaden 'the arena of institutions involved in guiding economic outcomes at diverse spatial scales' (ibid.: 55). In other words, it offers the prospect of a widespread social involvement in influencing the direction of the regional economy, through involvement in a diffuse and dispersed set of associations (or institutions).

In principle, linking associationalist approaches to the notion of institutional thickness should make it possible to operate 'in such a way that a region's ability to orchestrate becomes greater' (ibid.: 56). The process of developing institutions and associations through negotiation is, for Amin and Thrift, a vital element in providing the basis on which communities 'may develop sufficient potential for strategic action' (ibid.: 55). Amin and Thrift stress the democratizing potential of associations, both to overcome what they see as an existing democratic deficit within the institutions of economic governance and as a vital element in underpinning the development of progressive coalitions capable of delivering successful growth. Ensuring that associations operate in an open way is, for them, as important as fostering the existence of associations in the first place.

This way into the problem is potentially very exciting because it opens up alternative democratic agendas, as well as suggesting ways of developing wider economic agendas at regional level. But it raises problems of its own, too. One, which is recognized by Amin and Thrift (as well as Hirst, 1995) and is familiar from more orthodox pluralist writing is that all associations do not start out equal. Power is distributed differentially and this is bound to be reflected in the way in which associations operate. Within regions there can be little doubt that some groups and some interests will be more powerful than others. That is even more likely to be the case in the economic sphere, and particularly in the wake of neo-liberal restructuring which has effectively marginalized trade unions (in the south east still more than in other regions). Research on urban regimes and growth coalitions makes it clear that some agendas are likely to dominate at the expense of others. Indeed, the importance of non-decision-making in simply excluding some issues from business led agendas is hard to ignore (see Harding, 1995).

There is a related concern about the way in which groups are formed – who is able to join which. Amin and Thrift are clearly right to stress the importance of openness – and to look for ways of involving the 'whole' community – but it is no less clear that the balance is likely to operate in favour of some at the expense of others. Despite the optimism of some authors, it is just as likely that the process of forming associations and of coalition-building will be used to exclude some groups and some people from influence, instead of opening up the process. It is perhaps not surprising that some (such as Mansbridge, 1995; Levine, 1995) have seen associationalism as a form of neo-corporatism, nor that supporters of associationalism have chosen to avoid being tainted by that term.

One way of dealing with this might be to allow some scope for the state to intervene with a claim to have some overarching responsibility for managing relations between associations, possibly relying on a legitimacy drawn from electoral representation. Hirst (1994) sets out a fairly complex structure of regional assemblies with different forms of representation, mixing functional and electoral principles, and allocates them significant roles in terms of managing inequalities of power. Amin and Thrift follow a different line. They set out a programme for democratizing existing (state) networks and building alternative networks in civil society. In this way they hope to overcome what they see as a democratic deficit both in the operation of representative politics and in the potential power dynamics of associationalism.

A further problem with much of the debate about associationalism, from whatever perspective, is that, however attractive the prospect may be, the mechanism of transition is difficult to identify: in other words, it is difficult to see quite how to get from here to there. Leaving it to organic change seems likely to take too long – by the time it happens the neo-liberal agenda will already have been implemented. It is unclear who the agents of change will be. The outcome may be desirable, but the way of getting there is less clear. In his discussion of the arguments developed by Cohen and Rogers (1995), Offe notes that they fail to identify the 'reformist agents and political promoters' who can catalyse the necessary change. He asks if it is legitimate to conclude that associationalism is

> an arrangement of the greatest functional, but at the same time very limited normative appeal, which for this very reason is quite unlikely to be adopted in contexts in which it does not already find favourable conditions due to historical antecedents.
>
> (Offe, 1995: 126)

Paradoxically, despite the non-statist basis of the theory, it looks as if only state action can generate moves in this direction. Yet such moves would either help to undermine the position of the state (which makes the state an unlikely agent) or the autonomy of the associations would be threatened by the continuing intervention of the state. Although these comments apply particularly strongly to the formulation developed by Cohen and Rogers (1995), they also highlight broader difficulties. Associationalism, even in the form discussed by Amin and Thrift, looks more like an ideal type (or a utopian vision) than a convincing political model, except in so far as succeeds in justifying the self-activity of a range of groups in setting out to influence change.

It is also difficult to see quite how divisions between regions would be resolved in an associationalist system – and this is a point made forcibly by Amin and Thrift (1995). A 'successful' region is unlikely to sacrifice itself for a less successful one. Again this suggests that an associationalist agenda would have to be backed up by a more extensive package of state policy, at national and European level. In other words, associationalism needs to operate within a wider state framework rather than being seen as an alternative to that framework. The problem of the south east is not just a problem for the south east.

In the end neither the south east nor Britain's other regions benefited from the 'growth' of the 1980s, and it is this understanding which should provide the basis for the development of a new approach to regional policy within the UK. The regional and local initiatives outlined above need to be set within a wider policy framework which seeks to integrate the development of the south east into the development of the other regions of the UK in the context of a wider European focus. Instead of concentrating investment and development in the south east, if future growth is to be sustainable (or achievable) a wider understanding will be required, which encourages the diffusing of growth by imaginative means. Active, and positive, intervention will be required, so that – for example – it is not simply taken for granted that terminals for the Channel Tunnel need to be located in the south east, or so that there is a real commitment to developing the transport infrastructure to link the east coast ports to Wales and Ireland (a project supported by the EU, but so far given little strong support by the UK government).

The south east's growth has proved to be self-defeating because it has been rooted in continuing processes of uneven development and inequality, within Europe and the region itself as well as within the UK. The political economy of regional growth was based on an appeal to inequality as its defining characteristic. Within the Thatcherite rhetoric and imagery, this inequality was not so much perceived as a necessary evil on the way to wider success, but rather as a constitutive component of a society based on entrepreneurialism and reward for individual (and by implication regional) merit. Not only, however, does such an approach fail to acknowledge the complex and cumulative processes which generate inequality, but it fails to recognize the ways in which inequalities may themselves combine to undermine the growth which seems to be generated by them.

At present the south east tends to act as a powerful filter through which the UK's main linkages to the outside world seem to pass. It is important to stress that this is not simply the spatial outcome of the operation of some 'natural' or necessary global (market) processes. As we have seen, the relations which underlie it have been actively constructed, particularly since the mid-1970s, as part of a neo-liberal political project for the UK as a whole The interaction of politics, economics and geography has helped to produce the south east as 'growth region'. The dominant imaginary geographies of the south east are powerful ones rooted in material as well as discursive 'realities', but they need to be forcibly challenged if an alternative political project is to be successfully developed, for the sake of those living within the region as well as those living outside it. The success and failure of the south east as a 'growth' region in the 1980s and 1990s has forcibly highlighted the limitations of the political strategy which helped to define it. It is only by rejecting the search for a 'growth' region whose success will rescue us all and instead recognizing the need for more balanced – and more even – development, that we can begin to build a framework of positive regional policy which will also support and encourage localized initiatives.

It is, however, important not to underestimate the scale of the task which confronts anyone trying to challenge the current regional balance of power. The south east remains the home of the City, with its focus on global financial markets (rather than the UK or even Europe), and it remains the home of public power, too, with a major concentration of government and government departments. It is still the wealthiest part of the UK. In the past, despite the initiatives of the 1970s, it has proved difficult to question the pre-eminence of the south east, and, indeed, it has often proved easier to avoid directly confronting the issue. In the future there is a danger that, despite the failures of the 1980s, this will continue to be the case. But the south east is no longer the UK's great exception – it is no longer somehow exempt from the problems of economic restructuring and decline. On the contrary, it too is in an economically precarious position. This means that the opportunity for change is there. The time has come – indeed it is overdue – for the south east to give up its claims to superiority and to accept that it, too, needs to be integrated into a comprehensive regional policy capable of balancing the needs of different regions against each other. Otherwise there is little prospect for the generation of stable and sustained growth even in the south east.

6

SPACE, PLACE AND TIME

The argument of this book has focused on how to think about places and how 'thinking in terms of relations' – the nub of our approach – can throw new light on the nature of places and on the formation of their identities. Thinking in terms of relations forces you to think about conceptualization. It is simply not possible, from such a viewpoint, to take existing bounded regions/places as simply given to the analysis. This is not to say that such boundaries will never adequately define a region, nor that they can be assumed not to be important; rather, it is merely to stress that they should never be taken unquestioningly as adequate definitions.

There is, moreover, one supremely important aspect to this. Thinking in terms of relations has much to recommend it in social analysis generally, and as an approach it has been much written about and advocated, perhaps particularly by scholars writing from a feminist position. What we argue here is that it has an additional advantage in analyses which are – or aim to be – spatially sensitive. Thinking of a region as a pre-given bounded space is to think of it, first, only spatially. What is significant about thinking in terms of relations in a geographical analysis is that social relations are, integrally and inseparably, both social and spatial. On the one hand, our conceptualization of space itself is as social-relations-stretched-out. What happened in the 1980s was the restructuring of social relations in which a new space, or rather a new space–time, was produced. On the other hand, equally, there can be no social relation which does not have a spatial form.

Our approach should therefore be distinguished from Giddens's (1981, 1984) formulation of the notion of time–space distanciation. In part this is because our conceptualization of the spatiality of social relations aims to capture more than a spatial relation of distancing – our own formulation of social relations as 'stretched-out' might therefore also be seen as inadequate! Relations of juxtaposition and articulation may be equally important, as the previous chapters have illustrated. Further, our approach is different from that of Giddens because the notion of time–space distanciation is set, sometimes implicitly and sometimes explicitly, against an assumption that distanciation is a departure from a past in which social relations were spatially immediate. That formulation seems to us inadequate both because, we hold, social relations are always inevitably spatialized (that is, they have some kind of spatial form) and because Giddens' subsequent analysis seems to imply that the postulated immediacy (or even 'localness') erases all problems of communication and interpretation.

The approach adopted here, then, has the effect of thinking the social and the spatial together from the outset. The further implication is that space, and spatiality in general, is socially constructed. Moreover, that process of construction is constantly evolving. The spatialities of our lives are the product of continual 'negotiation', the outcome of the articulation of differentially powerful social relations. This applies both to the construction of spatial boundaries (of Buckinghamshire, of the Standard Region, or whatever) and to the spatialization of all other social relations. It is the recognition of the continuousness of this process of construction that has led us to think in terms of space–*time*. Any settlement of social relations into a spatial form is likely to be temporary. Some settlements will be longer lasting than others. The general lines of some elements of the dominance of London have broadly endured for centuries; the creation of Milton Keynes provoked a relatively sudden and sustained reorganization of spatialities over at least a half of England; the negotiation of the work:home boundary between the high-tech scientist and his partner may evolve from week to week. In principle, therefore, it is important at least to be aware of the essential temporality of all spatial figurations. It is in that sense that the-south-east-in-the-1980s was a distinct space–time.

The spaces, places and times of neo-liberalism

Integrating the social and the spatial enables a clearer appreciation of the social significance of spatial organization. The events of the 1980s changed not just the internal social structure of the south east (though they most certainly – as we have seen – did that), they also reworked the geography of those social relations both within and without the region. This reworked geography, moreover, was integral to the process – it was an essential element in the terms of the economic 'success' of the region in this period.

The free-market form of growth produced the south east as a space with a set of highly particular characteristics. Looking back over the analyses of the last few chapters, a number of these seem particularly clear. Some of them might equally well have emerged under different kinds of economic strategies, but their particular form, and the bundle of them together, is specific to neo-liberalism.

First, the restructuring of the region through the particular geography of free-market growth dynamics increased the separation of this part of the country from the rest. This is more than simply saying that 'the north–south divide' was widened during the period. This is certainly true and, as a serious exacerbation of a form of uneven development, as we have seen in Chapter 5, it had real implications for the sustainability of this form of growth. However, this (usual) way of thinking about the north–south divide, or uneven development more generally, concentrates on indices and levels of economic success. It is, to recall the terminology used in Chapter 2, to imagine regional space as an undulating surface. It is, however, space as social relations – the north–south divide conceptualized in terms of interregional relations – which is in question here. And what seems to have happened in the 1980s is that the intensity and the relative importance of the social relations linking the south east with the rest of the country was quite considerably diminished. In part this was the result of ongoing processes of globalization and the more general effect of spatial disarticulation within local areas to which this leads. It seems a fair bet to say that a map of the geography of economic relations 50 years ago, or even in the 1960s, would reveal patterns of local/regional interlinkage (although certainly not, even then, spatial 'coherence' or closedness) which today either no longer exist or are significantly weaker. The increasing internal complexity of the south east, therefore, goes hand in hand with an opening of the national economy and a decrease in the significance of interregional relations. This long ongoing process was accelerated and intensified during the 1980s, and it continues today. The global hegemony of the discourse and practice of 'free trade' was the essential term here. There was a deliberate and politically generated attempt to prize open local and regional economies. In the United Kingdom perhaps the most dramatic demonstration of this intent took place, precisely in the south east of England, with the abandonment of all exchange controls immediately on Mrs Thatcher's coming to power. This jettisoning of the power of the national state (deeply ironical in the context of apparently championing 'sovereignty' in the face of the European Union) would lead to major effects on the geography of social relations. Even more than it had been during its Empire (and informal empire) days the City became an enclave that primarily turned outwards away from the national economy and towards its deeper links into economic relations internationally.

Second, levels of inequality within the region increased – not only absolutely but also in relation to levels of inequality within other regions of the country. The south east, the region of greatest economic growth, became even more markedly also the region of greatest contrasts between rich and poor. Moreover, as we have tried to show, two particular characteristics of this inequality stand out. First, it was an inherent structural characteristic of this kind of growth itself. Second, it had its own, definite, geography. Chapters 3 and 4 have examined this in detail – the geographical and social divides of both exclusion and inclusion. On the one hand, the places and social groups structurally excluded from this form of growth; on the other hand, the poverties of inclusion – of those who service the growth – gets all the attention.

Third, this process has gone along with the production of new place identities and new social identities – a new symbolic social geography both further dividing the south east from the north of the country and reworking the internal socio-symbolic geography of the south east itself. The yuppies invading the City and the scientist-entrepreneurs capturing the new dynamic growth in Cambridge by no means entirely overthrew the hold over certain spaces exercised by the old elite. And in places such as Buckinghamshire their dominance was similarly maintained, although even here new money can always threaten to lower the tone – for wealth is not the same thing, in this language, as class. But in these places and in the south east more generally the project of free-market liberal economics disturbed – both socially and geographically – the hold of the previous social dominant. The very terms of the south east's regional dominance on which the growth of the 1980s was built were challenged by the free-market values and rhetoric of individualism and competitive enterprise which were central to the Thatcherite project. A first point that can be made in this regard, therefore, is that the south east which soared away from the rest of the country in economic terms in the 1980s was, even in terms of its ruling class, not quite the same south east which had previously been dominant. This was not simply the widening, in other words, of an already established inequality; it was a reworking of the class terms of interregional inequality. Effectively, this element of the class dimension of the north–south divide was restructured. As we shall see, it was not the only element of that long historical-national division to be refashioned.

As has also been argued, this 'social' restructuring took place along lines of ethnicity and gender as well as along those of class. The established white Englishness of the City, Cambridge and the old home counties gave a little in the face of, and was invaded by, an equally white thrusting masculinity of the new arrivals. The white ROSE remained white, but the kind of whiteness being constructed there was new. The new 'edge cities' do not on the whole have the ethnic diversity of the metropolis. This is smaller-scale, white urbanism. Moreover, and as if to emphasize the point, those towns within the outer reaches of the region which do have higher (that is, higher than ROSE) proportions of ethnic minority groups within their populations – towns such as Bedford, Luton, Slough – tend to be those which represent the higher water mark of a previous round of growth, based on (Fordist) mass-production manufacturing industry, and particularly on the car industry. They were not always entirely excluded by the growth of the 1980s but neither were they hot spots. This raises again the possibility of an ethnic dimension to the neo-liberal reproduction of uneven development, an issue which will be returned to below.

It is, moreover, bitterly ironic that the dominant ethos of the period of office of the first woman prime minister was utterly masculine. The rhetoric of thrusting individualistic workaholism was deeply male. The iconic figures we have examined most closely – the yuppie and the scientist-entrepreneur – were irretrievably masculine (which is absolutely not the same as saying that they were all men, though most of them were), but so too were many others, from the barrow boy to the estate agent to the practitioner of the new brand of macho management. The very discourses of the new economics had a masculinized erotic charge. In many a conversation with people working in newly privatized utilities, or even in local authorities, comments were made about the sexual form of the excitement in people suddenly able to deal in 'big Buckinghamshire', to talk in terms of millions of 'K', to refer to the hard reality of 'the bottom line'. It was a tedious but highly important aspect of what we might call the gendering of economic discourse.

It was not, of course, that economic discourse had not been gendered before. The classical or neo-classical figure of economic 'man' immediately puts that hypothesis to rest. Rather, once again, part of what is at issue is the changing *form* of social identities – here the changing form of the dominant masculinities. Moreover, the kinds of aggressive, 'thrusting', competitive and workaholic masculinities described in Chapter 4 have, as that chapter also argued, particularly deleterious effects on others – especially those who have to service them or try to live with them.

The growth dynamics of the 1980s – the ones which produced the new south east – were overwhelmingly dominated by varieties of this kind of masculinity (think of finance, of high technology, of the property market). In this it contrasts markedly with, for instance, the economic project of the 1960s. The social democratic project of Harold Wilson was by no means feminist in its impetus; it, too, for instance, revered the white heat of technology – though in a different form – and its emphasis on 'big is beautiful' entailed the glorification of masculine figures, Arnold Weinstock for one. But it did, none the less, have some interesting effects on gender relations, most particularly through the massive expansion of the welfare state. The growth of the south east in the 1980s was both predicated upon and entailed the growth and glorification of the kinds of masculinities we have examined here.

Moreover, the reworking of the imagery of the south east entailed the region's *identification* with these figures. This by no means implies that the south east was coded masculine in contradiction to a 'north' which was feminine. For much of Mrs Thatcher's mission, and indeed the mission of the rapidly rightward moving Labour Party, was to attack 'old' forms of identification around manufacturing industry, manual labour, and trade unions. And the caricature coding-figure for this part of the economy, and this part of the country, was the macho working-class trade unionist. Ridiculed during the whole decade, these figures were constantly derided for their lack of attention to new movements such as feminism. In itself, such a critique undoubtedly had substance. But what it omitted, and what was absolutely crucial, was that the 'new' economic figures which were to make the old one redundant were equally masculinist. One possible reading of the symbolic geography of the north–south divide during this period is of a battle between two different forms – equally objectionable – of masculinity.

Fourth, the dynamics of neo-liberal uneven development within this heartland of its success displayed particular characteristics. It is widely agreed that, over the last century, many branches of industry have become increasingly freed from locational ties to natural resources. It is also often argued that even geographical ties to labour skills in the old sense have been loosened. Moreover, technological changes in transport and communications have given many parts of the economy an enormously high degree of potential mobility. Finally, it is also often argued that questions of image and of symbolic status have grown dramatically in significance. One conclusion sometimes drawn from all this is that industry now has a high degree of 'footlooseness' in the sense of locational indifference. In one sense this is true: there are perhaps fewer potentially decisive economic differences between locations. If this is so, then, in principle, and ironically given the anti-interventionist tenor of the times, urban and regional policies designed to combat the inequities of uneven development should be more feasible than they have been in the past.

This, of course, has not happened. There have been no attempts through industry policy to ameliorate the impact of a geographical uneven development which worsened markedly under neo-liberal economic strategies. The dominant ethos ruled that the market knew best. However, precisely those same factors outlined above have brought about a particular dynamic of the reproduction of uneven development. In this dynamic, class, status, fashion and 'environment', can be all important. What emerges from this study of the south east is that areas of ethnic mix, areas of decline, areas with high unemployment and areas of physical decay will be avoided. Areas with characteristics the opposite of all these, areas with 'good' environmental credentials, will attract the growth. (Of course, intra-area levels of economic equality/inequality may also be significant depending upon the kind of growth in question.) This is emphatically true of some of the most significant growth dynamics examined in this book. The result is that areas which need growth less get it while areas in desperate need of new investment do not. The costs of congestion are piled up alongside the costs of decline. And – the deepest irony – this occurs for reasons which largely revolve around social and spatial antipathy and exclusion. It is little wonder that at all geographical scales, from the global to the very local, free-market economic strategies have resulted in increasing levels of geographical inequality.

Fifth and finally, these hard processes of social and spatial exclusion could lead either to the undermining of these economic strategies as untenable or to such levels of inequality as will only be containable by even more severe repression. On the global scale we can already see the confrontation between the levels of uneven development resulting from untrammelled free trade and free-market policies on the one hand and policies to restrict the free movement of peoples – international migration – on the other. The increasing severity of uneven development is part of what gives rise to the pressure for migration. Hirst and Thompson spell out some of the internal entanglements of this situation:

> Internationally open cultures and rooted populations present an explosive contradiction. The impoverished can watch 'Dallas'. They know another world is possible . . . A world of wealth and poverty, with appalling and widening differences in living standards between the richest and the poorest nations, is unlikely to be secure or stable.
>
> (Hirst and Thompson, 1996: 182)

Similar contradictions can be identified at other geographical scales. In the United Kingdom the widening of the north–south divide, within the different institutional setting of a national economy, led to such contradictions. Thus, in spite of the rhetoric of flexibility, the uneven development of the national economy in the 1980s in fact led to labour market rigidities. Unemployed people in the north were unable to move to the south to 'cool' the demand for labour there. The shortages of labour in the south east held back output. The escalation of house prices in the south east led, in a situation of national wage-bargaining, to wages being higher than they might otherwise have been (hence the political insistence on local wage bargaining – though in fact high wages were not the economy's main problem). Infrastructure in the north remained underused while more had to be provided in the overburdened south. Overall, this combination of overheating and underuse – the result of exacerbated geographical uneven development – enforced the earlier imposition of macro-economic restraints on growth than would otherwise have been necessary. Geographical inequality emerged as one of the classic vulnerabilities of neo-liberal growth.

Where does this leave the south east today? Well, for one thing, there is no social democratic wand which will wipe out at a stroke the social and geographical inequities or the deep cultural legacy of individualism and competition which have become more or less embedded in the social fabric of south east. As we have stressed throughout, it was not as if a particular form of growth was laid down intact across the south east at one moment in time. It is, of course, the case that the different components of growth came together in the 1980s, to produce the south east as a particular neo-liberal space, but – and this is important to stress – those different elements were the product of both long- and short-term changes at work across the region. They were products of different time-spans, the outcome of a set of differences in the pace and tempo of change. As such, many of the changes we have discussed here, in their different ways, have become emeshed in the social fabric of the south east, with some changes more entrenched than others, but together they represent the context which succeeding sets of relationships will have to negotiate.

From the viewpoint of the mid-1990s, for many social groups the grotesque inequalities and strident individualism of the past decade and a half may not be an acceptable price to pay for economic 'success'. Whatever kind of growth *is* sought in the south east, however, it will take place in the context of a profoundly neo-liberal region. Above all, in coming to terms with that legacy, it is the message of this book that an adequate understanding of the region and its futures can only come through a conception of places as open, discontinuous, relational and internally diverse. In short, regions are a construction in space–time: a product of a particular combination and articulation of social relationships stretched over space. To see them as anything less, is to settle for an inadequate understanding of contemporary regional geographies.

BIBLIOGRAPHY

Acker, J. (1990) 'Hierarchies, jobs, bodies: a theory of gendered organisation' in *Gender and Society*, 4: 139–58.

Allen, J. (1992) 'Services and the UK space economy: regionalization and economic dislocation', in *Transactions, Institute of British Geographers*, 17, 292–305.

Allen, J. (1997) 'Precarious work and shifting identities: contract labour in a global city' in A. Martens and M. Vervaeke (eds) *Polarisation sociale des villes européennes*, Lille: Anthropos.

Allen, J. and Henry, N. (1995) 'Growth at the margins: contract labour in a core region' in C. Hadjimichalis and D. Sadler (eds) *Europe at the Margins: New Mosaics of Inequality*, Chichester: John Wiley.

Allen, J. and Henry, N. (1996) 'Fragments of industry and employment: contract service work and the shift towards precarious employment' in R. Crompton, D. Gallie and K. Purcell (eds) *Changing Forms of Employment: Organisations, Skills and Gender*, London and New York: Routledge.

Allen, J. and Henry, N. (1997) 'Ulrich Beck's "Risk Society" at work: labour and employment in the contract service industries' in *Transactions, Institute of British Geographers*, 21: 180–96.

Allen, J. and Pryke, M. (1994) 'The production of service space' in *Environment and Planning D: Society and Space*, 12: 453–75.

Amin, A. and Thrift, N. (1992) 'Neo-Marshallian nodes in global networks' in *International Journal of Urban and Regional Research*, 16: 571–87.

Amin, A. and Thrift, N. (eds) (1994) *Globalisation, Institutions and Regional Development in Europe*, Oxford: Oxford University Press.

Amin, A. and Thrift, N. (1995) 'Institutional issues for the European regions: from markets and plans to socioeconomics and powers of assocation' in *Economy and Society*, 24, 1: 41–66.

Anderson, P. (1964) 'Origin of the present crisis' in *New Left Review*, 23: 11–52.

Barham, J. (1986) *Backstairs Cambridge*, Orwell: Ellison Editions.

Barlow, J. and Fielding, A. (1992) 'Population and migration trends', in I. Breugel (ed.) *The Rest of the South East: A Region in the Making. A Study for SEEDS*, London: School of Land Management and Urban Policy, South Bank University.

Barlow, J. and Savage, M. (1986) 'The politics of growth: cleavage and conflict in a Tory Heartland' in *Capital and Class*, 31: 156–82.

Barrett Brown, M. (1988) 'Away with all the great arches: Anderson's history of British Capitalism' in *New Left Review*, 167: 22–51.

Beck, U. (1992) *Risk Society: Towards a New Modernity*, London: Sage.

Begg, I.G. and Cameron G.L. (1988) 'High technology location and the urban areas of Great Britain' in *Urban Studies*, 25: 366–79.

Bendixson, T. and Platt, J. (1992) *Milton Keynes: Image and Reality*, Cambridge: Granta Editions.

Benington, J. (1994) *Local Democracy and the European Union: The Impact of Europeanisation on Local Governance*, London: Commission for Local Democracy.

Bhabha, H.K. (1994) *The Location of Culture*, London and New York: Routledge.

Breheny, M. and Congdon, P. (1989) *Growth and Change in a Core Region*, London: Pion.

Buxton, A. (1988) 'Where dishevelled dons are giving way to boffins with more BMWs than bikes' in *Guardian*, 5 September.

Cain, P.J. and Hopkins, A.G. (1986) 'Gentlemanly capitalism and British expansion overseas I. The old colonial system, 1688–1850' in *Economic History Review*, 2nd series, 39: 501–25.

Cain, P.J. and Hopkins, A.G. (1987) 'Gentlemanly capitalism and British expansion overseas II. New imperialism, 1850–1945' in *Economic History Review*, 2nd series, 40: 1–26.

CERN (1993) *Rethinking Urban Policy: Bulletin No.1*, Manchester: CLES European Research Network Ltd.

Charlesworth, J. and Cochrane, A. (1994) 'Tales of the Suburbs: the local politics of growth in the south east of England' in *Urban Studies*, 31, 10: 1723–38.

Charlesworth, J. and Cochrane, A. (1997) 'Anglicising the American dream: tragedy farce and the "post-modern" city' in S. Westwood and J. Williams (eds) *Imagining Cities: Scripts, Signs, Memory*, London and New York: Routledge.

Church, A., Cundell, I., Hebbert, M., McCoshen, U., Palmer, D., Pinto, R., Rainnie, A., Sellgin, J. and Wood, P. (1992) 'The south east' in P. Townsend and R. Martin (eds) *Regional Development in the 1990s: The British Isles in Transition*, London: Jessica Kingsley.

Cloke, P., Phillips, M. and Thrift, N. (1995) 'The new middle classes and the social constructs of rural living' in T. Butler and M. Savage (eds) *Social Change and the Middle Classes*, London: UCL Press.

Coakley, J. (1984) 'The internationalisation of bank capital' in *Capital and Class*, 23: 107–20.

Coakley, J. (1992) 'London as an international finance centre' in L. Budd and S. Whimster (eds) *Global Finance and Urban Living*, London and New York: Routledge.

Coakley, J. and Harris, L. (1983) *The City of Capital: London's Role as a Financial Centre*, Oxford: Basil Blackwell.

Cochrane, A. (1993a) 'Looking for a European Welfare State' in A. Cochrane and J. Clarke (eds) *Comparing Welfare States: Britain in an International Context*, London: Sage.

Cochrane, A. (1993b) *Whatever Happened to Local Government?* Buckingham: Open University Press.

Cochrane, A. (1994) 'Beyond the nation state? Building Euro-regions' in U. Bullmann (ed.) *Die Politik der Dritten Ebene. Regionen in Europa der Union*, Baden-Baden: Nomos.

Cochrane, A., Seavers, J. and Sarre, P. (1996) *Maximising Participation in the Labour Market*, Milton Keynes: Report to Milton Keynes Economic Partnership.

Cockburn, C. (1983) *Brothers: Male Dominance and Technological Change*, London: Pluto Press.

Cockburn, C. (1991) *In the Way of Women: Men's Resistance to Sex Equality in Organisations*, Basingstoke: Macmillan.

Cohen, J. and Rogers, J. (1995) *Associations and Democracy: The Real Utopian Project, Vol. 1* (ed. E. Olin Wright), London: Verso.

Collinson, D. and Hearn, J. (1994) 'Naming men as men: implications for work, organisations and management' in *Gender, Work and Organisation*, 1, 1: 2–22.

Cooke, P. (ed.) (1989) *Localities: The Changing Face of Urban Britain*, London: Unwin Hyman.

Cooke, P., Moulaert, F., Swyngedouw, E., Weinstein, D. and Wells, P. (1992) *Towards Global Localisation: The Computing and Communications Industries in Britain and France*, London: UCL Press.

Court, G. and McDowell, L. (1993) 'Serious trouble? Financial services and structural change', *Occasional Paper Series 4, The South East Programme*, Milton Keynes: The Open University.

Crang, P. and Martin, R.L. (1991) 'Mrs Thatcher's vision of the "new Britain" and the other sides of the Cambridge phenomenon' in *Environment and Planning D: Society and Space*, 9: 91–116.

Cross, M. and Waldinger, R. (1993) 'Migrants, minorities and the ethnic division of labour' in S.S. Fainsten, I. Gordon and M. Harloe (eds) *Divided Cities: New York and London in the Contemporary World*, Oxford: Basil Blackwell.

Cutler, T., Hallam, C., Williams, J. and Williams, K. (1989) *The Struggle for Europe*, Oxford: Berg.

Deakin, S. (1992) 'Labour law and industrial relations' in J. Michie (ed) *The Economic Legacy 1979–1992*, London: Academic Press.

Deakin, S. and Wilkinson, F. (1992) 'European integration: the implications for UK policies on labour supply and demand' in E. McLaughlin (ed.) *Understanding Unemployment: New Perspectives on Active Labour Market Policies*, London and New York: Routledge.

Dorling, D. (1993) 'The spread of negative equity' in *Findings*, Joseph Rowntree Foundation.

Directorate-General for Regional Policy (1991) *Europe 2000: Outlook for the Development of the Community's Territory*, Luxembourg: European Commission.

Duncan, S. (1989) 'What is a locality?' in R. Peet and N. Thrift (eds) *New Models in Geography: The Political Economy Perspective*, London: Unwin Hyman.

Dunford, M. and Kafkalis, G. (eds) (1994) *Cities and Regions in the New Europe: The Global–Local Interplay and Spatial Development Strategies*, London: Belhaven.

Esping-Anderson, G. (1993) *Changing Classes: Stratification and Mobility in Post-industrial Societies*, London: Sage.

Fielding, A. J. (1992) 'Migration and social mobility: south east England as an escalator region' in *Regional Studies*, 26: 1–15.

Forrest, R. and Gordon, D. (1993) *People and Places: A 1991 Census Atlas of England*, Bristol: School for Advanced Urban Studies.

Gamble, A. (1988) *The Free Economy and the Strong State: The Politics of Thatcherism*, Basingstoke: Macmillan.

Garreau, J. (1991) *Edge City: Life on the New Frontier*, New York: Doubleday.

Giddens, A. (1984) *The Constitution of Society*, Cambridge: Polity Press.

Giddens, A. (1981) *A Contemporary Critique of Historical Materialism Vol. 1: Power, Property and the State*, Basingstoke: Macmillan.

Gilroy, P. (1987) *There Ain't No Black in the Union Jack*, London: Hutchinson.
Gilroy, P. (1993) *The Black Atlantic: Modernity and Double Consciousness*, London and New York: Verso.
Gorz, A. (1989) *Critique of Economic Reason*, London and New York: Verso.
Gregson, N. and Lowe, M. (1994) *Servicing the Middle Classes: Class, Gender and Waged Domestic Labour in Contemporary Britain*, London and New York: Routledge.
Hall, C. (1992) *White, Male and Middle Class: Explanations in Feminism and History*, Cambridge: Polity Press.
Hall, S. (1990) 'New ethnicities' in J. Donald and A. Rattansi (eds) *'Race', Culture and Difference*, London: Sage.
Hall, S. (1991) 'The local and the global: globalisation and ethnicity' in A.D. King (ed.) *Culture, Globalisation and the World-system*, Basingstoke: Macmillan.
Hamnett, C. (1985) 'Urban land use: change and conflict' in *Changing Britain, Changing World: Geographical Perspectives*, Milton Keynes: The Open University.
Hamnett, C. (1992) 'The geography of housing wealth and inheritance in Britain' in *The Geographical Journal*, 158: 307–21.
Hamnett, C. (1994) 'Restructuring housing finance and the housing market' in Corbridge, S., Martin, R. and Thrift, N. (eds) *Money, Power and Space*, Oxford: Blackwell.
Hamnett, C. and Seavers, J. (1994a) 'The geography of house prices and house price inflation in the south east of England' in *Occasional Paper Series 12, The South East Programme*, Milton Keynes: The Open University.
Hamnett, C. and Seavers, J. (1994b) 'A step up the ladder? Home ownership housing careers in the south east of England' in *Occasional Paper Series 15, The South East Programme*, Milton Keynes: The Open University.
Hamnett, C. and Seavers, J. (1995) 'Winners and losers: the distribution of capital gains and losses from home ownership in the south east of England' in *Occasional Paper 16, The South East Programme*, Milton Keynes: The Open University.
Harding, A. (1995) 'Elite theory and growth machines' in D. Judge, G. Stoker and H. Wolman (eds) *Theories of Urban Politics*, London: Sage.
Harvey, D. (1985) 'The geopolitics of capitalism' in D. Gregory and J. Urry (eds) *Social Relations and Spatial Structures*, Basingstoke: Macmillan.
Hay, C. (1995) *Re-stating Social and Political Change*, Buckingham: Open University Press.
Henry, N. and Massey, D. (1995) 'Competitive times in high/tech' in *Geoforum* 26, 1: 49–64.
Hirst, P. (1994) *Associative Democracy: New Forms of Economic and Social Governance*, Cambridge: Polity Press.
Hirst, P. (1995) 'Can secondary associations enhance democratic governance?' in J. Cohen and J. Rogers *Associations and Democracy: The Real Utopian Project, Vol. 1* (ed. E. Olin Wright), London: Verso.
Hirst, P. and Thompson, G. (1996) *Globalisation in Question*, Cambridge: Polity Press.
Hudson, R. (1988) 'Labour market changes and new forms of work in "old" industrial regions' in D. Massey and J. Allen (eds) *Uneven Re-development: Cities and Regions in Transition*, London: Hodder and Stoughton.
Hutton, W. (1995) *The State We're In*, London: Cape.
Ingham, G. (1984) *Capitalism Divided? The City and Industry in British Social Development*, Basingstoke: Macmillan.
Ingham, G. (1989) 'Commercial capital and British development: a reply to Michael Barrett Brown' in *New Left Review*, 172: 45–65.

Jessop, B. (1994) 'The transition to post-Fordism and the Schumpeterian workfare state' in R. Burrows and B. Loader (eds) *Towards a Post-Fordist Welfare State?*, London and New York: Routledge.

Jessop, B., Bennett, K. and Bromley, S. (1990) 'Farewell to Thatcherism? Neo-liberalism and "New Times"' in *New Left Review*, 179: 81–102.

Jessop, B., Bennett, K., Bromley, S. and Ling, T. (1989) *Thatcherism: A Tale of Two Nations*, Cambridge: Polity.

Jessop, B. and Stones, R. (1992) 'Old City and new times: economic and political aspects of deregulation' in L. Budd and S. Whimster (eds) *Global Finance and Urban Living*, London and New York: Routledge.

Knox, P. (1993) *The Restless Urban Landscape*, Englewood Cliffs, NJ: Prentice Hall.

Laclau, E. (1990) *New Reflections on the Revolution of Our Time*, London and New York: Verso.

Lee, C.H. (1984) 'The service sector, regional specialisation, and economic growth in the Victorian economy' in *Journal of Historical Geography*, 10: 139–55.

Lee, C.H. (1986) *The British Economy since 1700: A Macro-economic Perspective*, Cambridge: Cambridge University Press.

Levine, A. (1995) 'Democratic corporatism &/versus socialism' in J. Cohen and J. Rogers *Associations and Democracy: The Real Utopian Project, Vol. 1* (ed. E. Olin Wright), London: Verso.

Leys, C. (1985) 'Thatcherism and British manufacturing: a question of hegemony' in *New Left Review*, 151: 5–25.

Leyshon, A. and Thrift, N. (1993) 'The restructuring of the UK financial services industry in the 1990s: a reversal of fortune?' in *Journal of Rural Studies*, 9: 223–41.

Leyshon, A., Thrift, N. and Daniels, P.W. (1987) 'The urban and regional consequences of the restructuring of world financial markets: the case of the City of London', *Working Paper on Producer Services*, 4, University of Bristol/Portsmouth.

Lloyd, J. (1986) 'A trinity of trading, growing and making' in *Financial Times*, 17 March.

Lock, D. (1989) *Riding the Tiger*, London: Town and Country Planning Association.

Lovering, J. (1991) 'The changing geography of the military industry in Britain' in *Regional Studies* 25: 279–93.

McDowell, L. and Court, G. (1994a) 'Gender divisions of labour in the post-Fordist economy: the maintenance of occupational sex segregation in the financial services sector' in *Environment and Planning* A, 26: 1397–418.

McDowell, L. and Court, G. (1994b) 'Missing subjects: gender, power, and sexuality in merchant banking' in *Economic Geography*, 70, 2: 229–51.

McDowell, L. and Court, G. (1994c) 'Performing work: bodily representations in merchant banks' in *Environment and Planning D: Society and Space*, 12: 727–50.

McDowell, L. and Court, G. (1995) 'Body work: heterosexual gender performances in City workplaces' in D. Bell and G. Valentine (eds) *Mapping Desire*, London and New York: Routledge.

Mansbridge, J. (1995) 'A deliberative perspective on neo corporatism' in J. Cohen and J. Rogers *Associations and Democracy: The Real Utopian Project, Vol. 1* (ed. E. Olin Wright), London: Verso.

Marsden, J., Flynn, A., Murdoch, J., Lowe, P. and Munton, R.J. (1996) *Constructing the Countryside: An Approach to Rural Development*, London: UCL Press.

Martin, R. (1988) 'Industrial capitalism in transition: the contemporary reorganization of the British space-economy' in D. Massey and J. Allen (eds) *Uneven Re-development: Cities and Regions in Transition*, London: Hodder and Stoughton.

Massey, D. (1984) *Spatial Divisions of Labour: Social Structures and the Geography of Production*, London and Basingstoke: Macmillan.

Massey, D. (1988a) 'A new class of geography' in *Marxism Today*, May: 12–17.

Massey, D. (1988b) 'What's happening to U.K. manufacturing?' in J. Allen and D. Massey (eds) *The Economy in Question*, London: Sage.

Massey, D. (1995a) 'Masculinity, dualisms and high technology' in *Transactions, Institute of British Geographers*, NS 20: 487–99.

Massey, D. (1995b) *Spatial Divisions of Labour: Social Structures and the Geography of Production* (2nd edn), London and Basingstoke: Macmillan.

Massey, D. and Allen, J. (1995) 'High-tech places: poverty in the midst of growth' in C. Philo (ed.) *Off the Map: The Social Geography of Poverty in the UK*, London: CPAG.

Massey, D., Quintas, P. and Wield, D. (1992) *High-tech Fantasies: Science Parks in Society, Science and Space*, London and New York: Routledge.

Mohan, J. (1988a) 'Spatial aspects of health-care employment in Britain: aggregate trends' in *Environment and Planning A*, 20: 7–23.

Mohan, J. (1988b) 'Restructuring, privatization and the geography of health care provision in England, 1983–87' in *Transactions, Institute of British Geographers*, 13: 449–65.

Mohan, J. (1989) *The Political Economy of Thatcherism*, London and Basingstoke: Macmillan.

Mohan, J. (1995a) 'Missing the boat: poverty, debt and unemployment in the south east' in C. Philo (ed) *Off The Map: The Social Geography of Poverty in the UK*, London: CPAG.

Mohan, J. (1995b) *A National Health Service? The Restructuring of Health Care in Britain since 1979*, London: Macmillan.

Moore, B., Rhodes, J. and Tyler, P. (1986) *The Effects of Government Regional Policy*, London: HMSO.

Morgan, K. (1994) 'Innovating by networking: new models of corporate and regional development', in M. Dunford and G. Kafkalis (eds) *Cities and Regions in the New Europe: The Global–Local Interplay and Spatial Development Strategies*, London: Belhaven.

Morgan, K. and Sayer, A. (1988) *Microcircuits of Capital: Sunrise Industry and Uneven Development*, Cambridge: Polity.

Morley, D. and Robins, K. (1995) *Spaces of Identity: Global Media, Electronic Landscapes and Cultural Boundaries*, London: Routledge.

Murdoch, J. and Marsden, T. (1994) *Reconstituting Rurality: The Changing Countryside in an Urban Context*, London: UCL Press.

Murray, R. (1989) *Crowding Out: Boom and Crisis in the South East*, Stevenage: SEEDS Publications.

Murray, R. (1992) 'Europe and the new regionalisms' in M. Dunford and G. Kafkalis (eds) *Cities and Regions in the New Europe*, London: Belhaven.

Offe, C. (1995) 'Some skeptical considerations on the malleability of representative institutions' in J. Cohen and J. Rogers *Associations and Democracy: The Real Utopian Project, Vol. 1* (ed. E. Olin Wright), London, Verso.

Peck, J.A. and Tickell, A. (1992) 'Local modes of social regulation? Regulation theory, Thatcherism and uneven development' in *Geoforum*, 23, 3: 347–63.

Pryke, A. (1991) 'An international city going "global": spatial change and office provision in the City of London' in *Environment and Planning: Society and Space*, 9, 2: 197–222.

Pryke, M. (1994) 'Looking back on the space of a boom: redeveloping spatial matrices in the City of London' in *Environment and Planning* A, 26: 235–64.

Putnam, R.D. (1993) *Making Democracy Work*, Princeton, NJ: Princeton University Press.

Rajan, A (1990) *Capital People: Skills Strategies for Survival in the Nineties*, London: The Industrial Society.

Rubenstein, W.D. (1977) 'Wealth, elites and the class structure of modern Britain' in *Past and Present*, 76: 9–126.

Rubenstein, W.D. (1981) *Men of Property: The Very Wealthy in Britain since the Industrial Revolution*, London: Croom Helm.

Rubenstein, W.D. (1994) *Capitalism, Culture and Decline in Britain*, London and New York: Routledge.

Rubenstein, W.D. (1987) *Elites and the Wealthy in Modern British History*, Brighton: Harvester.

Rustin, M. (1994) 'Incomplete Modernity: Ulrich Beck's *Risk Society*' in *Radical Philosophy*, 67: 3–12.

Sarre, P. and Seavers, J. (1995) 'Changing reasons for owner occupation' in *Occasional Papers 17, The South East Programme*, Milton Keynes: The Open University.

Saunders, P. (1990) *A Nation of Home Owners*, London: Unwin Hyman.

Savage, M., Barlow, J., Dickens, P. and Fielding, T. (1992) *Property, Bureaucracy and Culture: Middle Class Formation in Contemporary Britain*, London and New York: Routledge.

SEEDS (1987) *North–South Divide*, Stevenage: South East Economic Development Strategy.

Sibley, D. (1995) *Geographies of Exclusion*, London and New York: Routledge.

Smith, D. (1989) *North and South: Britain's Economic, Social and Political Divide*, Harmondsworth: Penguin.

Smith, D. (1993) *From Bust to Boom: Trial and Error in British Economic Policy*, Harmondsworth: Penguin Books.

Stubbings, F.H. (1991) *Bedders, Bulldogs and Bedells: A Cambridge ABC*, Cambridge: Emmanuel College.

Thompson, P. and McHugh, D. (1990) *Work Organisations*, Basingstoke: Macmillan.

Thrift, N. (1989) 'Images of social change' in C. Hamnett, L. McDowell and P. Sarre (eds) *The Changing Social Structure*, London: Sage.

Thrift, N. (1996) 'The moneyed spaces of the 1980s Britain: class, consumption and culture' in *Working Paper on Producer Services*, 33, University of Bristol/Liverpool.

Thrift, N. and Leyshon, A. (1992) 'In the wake of money: the City of London and the accumulation of value' in L. Budd and S. Whimster (eds) *Global Finance and Urban Living*, London and New York: Routledge.

Thrift, N., Leyshon, A. and Daniels, P. (1987) 'Sexy greedy: the new international financial system, the City of London and the South East of England' in *Working Paper on Producer Services*, University of Bristol/Liverpool.

Townsend, P., with Corrigan, P. and Kowerzik, U. (1987) *Poverty and Labour in London*, London: Low Pay Unit.

Turner, B.S. (1994) *Orientalism, Postmodernism and Globalism*, London and New York: Routledge.

Waldinger, R. (1992) 'Native blacks, new immigrants and the post-industrial transformation of New York' in M. Cross (ed.) *Ethnic Minorities and Industrial Change in Europe and North America*, Cambridge: Cambridge University Press.
Webster, A.J. (1989) 'Privatisation of public sector research: the case of a plant breeding institute' in *Science and Public Policy*, 16, 4: 224–32.
Weiner, M.J. (1981) *English Culture and the Decline of the Industrial Spirit, 1850–1980*, Harmondsworth: Penguin.
Zukin, S. (1991) *Landscapes of Power: From Detroit to Disney World*, Berkeley and Los Angeles: University of California Press.

The South East Programme: Occasional Paper Series – The Open University

OP1 *The nature of a Growth Region: The Peculiarity of the South East*
John Allen
November 1992

OP2 *Something New, Something Old: A Sketch of the Cambridge Economy*
Doreen Massey and Nick Henry
November 1992

OP3 *Notes on Housing and Class Formation: Theoretical Debates and Practical Measures*
Phil Sarre
December 1992

OP4 *Serious Trouble? Financial Services and Structural Change*
Gill Court and Linda McDowell
January 1993

OP5 *The Missing Subject in Economic Geography*
Linda McDowell and Gill Court
January 1993

OP6 *The Local Politics of Growth in a 'Growth' Region*
Julie Charlesworth and Allan Cochrane
May 1993

OP7 *Geography, Gender and High Technology*
Doreen Massey
December 1993

OP8 *Competitive Times in High Tech*
Nick Henry and Doreen Massey
December 1993

OP9 *Tragedy, Farce and the Post-modern City: The Case of Milton Keynes*
Julie Charlesworth and Allan Cochrane
April 1994

OP10 *Scientists, Transcendence and the Work/home Boundary*
Doreen Massey
April 1994

OP11 *Divisions in Labour: Contract Services and the Emergence of a New Employment Regime*
John Allen and Nick Henry
April 1994

OP12 *The Geography of House Prices and House Price Inflation in the South East of England in the 1980s*
Chris Hamnett, Beverley Mullings and Jenny Seavers
November 1994

OP13 *A Stroke of the Chancellor's Pen: The Social and Regional Impact of the Conservatives' 1988 Higher Rate Tax Cuts*
Chris Hamnett
April 1994

OP14 *Reflections on Interviewing Experiences*
edited by Julie Charlesworth
November 1994

OP15 *A Step up the Ladder? Home Ownership Careers in the South East of England*
Chris Hamnett and Jenny Seavers
November 1994

OP16 *Winners and Losers: The Distribution of Capital Gains and Losses from Home Ownership in the South East of England*
Chris Hamnett and Jenny Seavers
July 1995

OP17 *Changing Reasons for Owner Occupation*
Phil Sarre and Jenny Seavers
September 1995

INDEX

academic establishment 76–9, 95–7
Acker, J. 94
Allen, J.: 'downside' of growth 25, 28, 80; effects of 'flexible' labour market 113; flexible working hours 82; growth and changing nature of jobs in the security industry 103; inequality as form of growth 79
'American' working practices 92
Amin, A. 68, 128–9; associationalism 132, 133, 134, 135; institutional thickness 125–6
Anderson, P. 16
anti-enterprise culture 28, 97
anti-growth stance 87–9
'anti-industrial spirit' 16
associationalism 132–5

Bank of England 77
Barham, J. 81
Barlow, J. 67
Barrett Brown, M. 16
Beck, U. 112
Begg, I.G. 22
Bendixson, T. 86
Benington, J. 132
Bennett, K. 15, 26
Berkshire 67
Bhabha, H.K. 107
'Big Bang' reforms 14, 75
Blaijorism 117
boundaries 54–5; *see also* openness; 'doily' metaphor; relational approach
Breheny, M. 10, 52–3, 66
British Household Panel Survey 109
British Merchant Banking and Securities Houses Association 75
Bromley, S. 15, 26
Buckinghamshire 66–7; South Buckinghamshire 55, 83–4, 87–9
Buxton, A. 77

Cain, P.J. 92
Cambridge 56, 61, 66, 68, 72, 91; globalization, masculinity, class and space 73–82
Cambridge Phenomenon 76, 78
Cameron, G.L. 22
CERN 118
Channel Tunnel 124, 135
Charlesworth, J. 71, 83, 86
Chatham Maritime 71
City of London 16, 52, 91, 121; challenge to established middle class 28; commercial property market 21–2; ethnicity 105–6; finance as growth mechanism 18–19; globalization, masculinity, class and space 73–82; international financial links 49, 51, 68, 75–6, 139; new middle-class identities 92–5
city-states 66
civil service 11, 24
class: anti-growth resistance 88–9; globalization, masculinity, space and 73–82; middle class *see* middle class; regional class alliances 57; restructuring and inequality 139–40; 'servile' class 27; *see also* social polarization
cleaning services 100–3, 106–7
Cloke, P. 58
Coakley, J. 18
coalition-building 133–5
Cochrane, A.: agencies of growth 131; European social policy 128; heritage

theme 71; juxtaposition of place identities 83, 86, 87
Cockburn, C. 102, 105
Cohen, J. 134, 135
coherence, structured 57–8, 60
Collinson, D. 102
commerce 16
commercial property 21–2, 120
Committee of the Regions 130
commuting 122–3
competition 143; EU policies and 128; *see also* individualism
Congdon, P. 10, 52–3, 66
consumption: boom 4, 14–15, 108, 119–20; 'greater City' 35, 43; growth dynamic 19–21, 34–5
contract service industries *see* service industries
Cooke, P. 22, 60
core region 10, 11, 52–3
corporatism: neo-corporatism 134; regional 132
Corrigan, P. 25
cost: competing on 120–1; minimization 72
council housing 108, 111
counter-regional subsidies 23–4
Court, G. 75, 93
Crang, P. 81
credit 14–15, 119–20; growth dynamic 19–21, 34–5
credit-card debt 20
Cross, M. 106
cultural dominance 16
culture: anti-enterprise culture 28, 97; placed 29
customary elites 16, 24, 28–9; *see also* academic establishment, City of London, 'gentlemanly' capitalism
Cutler, T. 127

Daniels, P.W. 19, 74, 93
Deakin, S. 100
dealers, financial 92–5
debt *see* credit
decentralization 66, 75–6
decline 70, 72–3
defence industries 22, 35
deregulation 129; financial 14–15, 23; internal fragility and 123–5; narratives of growth 13–15
Development Corporation 85, 86
development planning 66–7, 83–5, 132
discontinuous regions 55–6, 70–3
discourse of dominance 10–13
docklands 124
'doily' metaphor, of the region 55–6, 68, 69, 70
domestic labour 27
dominance: narratives of growth 13–17; representation through growth 10–13
Dorling, D. 109
Duncan, S. 34
Dunford, M. 128

earnings *see* incomes
economic policy 35
edge cities 58, 67, 84, 140
employment: financial services 19, 35, 43, 51; 'greater City' 35, 43; high technology 22, 35, 44; social polarization 26–8
English Estates 71
Englishness 29; white 29–31, 105–7, 140
Enterprise Zone 71
entrepreneurial dynamism 12
entrepreneurial knowledge 97–9
entrepreneurial masculinity 92–5, 96
equilibration processes 73
Esping-Anderson, G. 27–8
'Essex person' 31
establishment 16, 24; challenge of Thatcherism 28–9; new middle class vs 91–9
ethnic niches 106–7
ethnicity 33, 90–1, 105–7, 140
Eurobond market 18
Euro-Keynesianism 128
Euromarkets 18
'Europe, pull of' 11, 12, 13
Europe of the Regions 128
'Europe 2000' strategy 121
European Union (EU) 55, 121–2; regional policy 127–9; regionally based politics 129–30, 131–2
Euro-Thatcherism 128
exchange controls 14, 18, 139
exclusion 71–3; discontinuous regions 55–6; ethnicity 105–7; poverty and 26; service ranks 100–5; spaces of 99–107
exceptionalism 16–17

federal fiscalism 128
Fielding, A.J. 23, 119
finance 74; employment 19, 35, 43, 51; identity of the City of London 79–80;

international links 49, 51, 68, 75–6, 139; mechanism of growth 18–19, 50–1; narratives of growth 14–15; new middle class 74–5, 92–5; *see also* City of London
financial deregulation 14–15, 23
Financial Services Act 1986 19
Fordism 33, 100; 'post-Fordist' men 100–5
foreign exchange market 18
Forrest, R. 35
fragility: growth 118–22; internal of a 'growth' region 122–6
'full–time' bundles of work 102

Gamble, A. 28–9
Garreau, J. 58, 66, 67
gender 140–1; and contract cleaning 100–3; earnings 34–5, 36, 37; *see also* masculinity, women
'gentlemanly' capitalism 12, 16; Thatcherite challenge 24–5, 28–9, 92–3
Germany 130–1
Giddens, A. 137
Gilroy, P. 105, 107
globalization of finance 18–19; international financial links 49, 51, 68, 75–6, 139; masculinity, class and space 73–82
Glover, T.R. 76
Gordon, D. 35
Gorz, A. 26–7
government expenditure 35; counter-regional subsidies 23–4; housing 47; regional spending per capita 46; roads and transport 45; trade and industry 48
government intervention/actions 23–4; associationalism 134–5; internal fragility of a 'growth' region 124–5; and national distributions 35–43; power relations 52
Greater London Council 125
Gregson, N. 27
group identities 90–107; City players 92–5; ethnicity 105–7; high-tech elites 95–9; service industries 100–5
growth 117–18; components of 4, 18–25; forms of 25–31; 'free-market' 2–3; 'holes' in 55–6, 68, 69, 70–3; 'hot spots' of 56, 66, 68, 69, 126; narratives of 13–17; nature of 56; representing dominance through 10–13; resistance to growth 83–4, 87–9; unequal and fragile 118–22
growth mechanisms 3–4, 17, 18–25; and discontinuous regions 70–3; 'hot spots' 66; national distributions and 34–43; *see also under individual mechanisms*
'growth' region 9–10, 136; internal fragility 122–6

Hall, C. 29
Hall, S. 105, 107
Hamnett, C.: boom and slump in property markets 20, 21, 108, 109; effects of financial deregulation 23; new share ownership 52
Harding, A. 134
Harris, L. 18
Harvey, D. 57
Hay, C. 117
Hearn, J. 102
Henry, N.: 'downside' of growth 25, 28, 80; effects of 'flexible' labour market 113; 'ever-present' culture of scientific entrepreneurs 98; flexible working hours 82; growth and changing nature of jobs in the security industry 103
Hertfordshire 70
high incomes: distribution of 35, 38, 39, 40
high technology 22–3, 52, 66; Cambridge Phenomenon 76–7; elites 95–9; employment 22, 35, 44; *see also* Cambridge
Hirst, P. 133, 134, 142
history 9, 53; dominance of growth 10–13; social inheritance 17
'holes', in growth 55–6, 68, 69, 70–3
Holford Report 78
'home counties' 33
homeownership 108–11, 113
honorary men 94
Hopkins, A.G. 92
'hot spots', of growth 56, 66, 68, 69, 126
hours of work, long 95, 98, 104–5
house prices 21, 35, 42
housing 87, 112–13; expenditure on 47; fragile growth 119–20, 124; identity-formation and contestation 107–11
Hudson, R. 35
Hutton, W. 120, 124

IBM 77

identity/identities 141; construction in Milton Keynes 84–6; dislocated 91; group *see* group identities; housing and 107–11; othering 82–3; of places 67–8; Thatcherism's challenge to white Englishness 29–31
image 71–3; Cambridge 78–9; Milton Keynes 84–6
imaginary geographies 2, 74
incomes 143; distribution of high incomes 35, 38, 39, 40; 'greater City' 35, 43, 51; regional change in earnings 35, 41; weekly earnings 34–5, 36, 37
individualism 2–3, 9, 143
industry: 'anti-industrial spirit' 16; government expenditure on 48; locational indifference 141–2; modernization 15; *see also* manufacturing
inequality/inequalities 3, 4–5, 89, 135; City of London and Cambridge 79; consumption as growth mechanism 19–20; neo-liberal growth 72–3, 139, 142–3; structural 26; 'two nation' politics 25–6; unequal growth 118–22
infrastructural investment 124
Ingham, G. 16
innovation 78
institutional thickness 125–6, 132–3
integration, economic 127–8
interdependence 54, 58, 66–7, 129; othering and 82–9; *see also* linkages
intermediate institutions, networks of 133–5
internal fragility 122–6
international migration 142
International Monetary Fund (IMF) 2
intra-regional variations 56
investment: financial institutions 19; foreign direct investment 77; public sector 23–4, 124; reason for house purchase 110; rounds of 70, 72–3

Jessop, B. 15, 19, 29, 117; 'two nation' strategy 25, 26, 118

Kafkalis, G. 128
Kent County Council 131
knowledge, entrepreneurial 97–9
Knox, P. 67
Kowarzik, U. 25

labour market: ethnicity and 105–7; rigidities 143; service sector 100–5; skilled labour shortages 119; uncertainties 112–13; *see also* employment, unemployment
Laclau, E. 91
Lawson, N. 14, 20, 21, 108
Lee, C.H. 16, 19–20
Levine, A. 134
Leys, C. 24
Leyshon, A. 15, 19, 74, 93
liberalization 13–15
linkages 54; international links 49, 51, 68, 75–6, 139
Lloyd, J. 77
local authorities: links with Europe 131–2; Medway Towns 70–1; South Buckinghamshire and anti-growth lobby 88–9
Local Enterprise Companies 131
localities projects 58–60
locational indifference 141–2
Lock, D. 132
London 138; City of *see* City of London; ethnicity 105–7; 'greater City' 35, 43; 'London region' 65–73; othering and interdependence 83; relation to outer South East 58; role in South East's dominance 11
long hours 95, 98, 104–5
Lovering, J. 22, 35
Lowe, M. 27
Luton 70

M25 124
machinery 100–1
machismo 93, 95
Mansbridge, J. 134
manufacturing 33; dominance of commerce 16; Wilson modernization strategy 15, 53–4; *see also* industry
Marsden, J. 67
Marsden, T. 67, 84
Martin, R. 35, 81
masculinity 91–105, 140–1; contract services 100–5; entrepreneurial 92–5, 96; globalization, class, space and 73–82; high-tech elites 95–9
Massey, D. 15, 25, 77, 78, 124; contract cleaning 82; high technology 22, 23, 98; poverty 79; social polarization 26; uneven development 119
McDowell, L. 75, 93

McHugh, D. 101
Medway Towns 56, 70–1
middle class: City players 92–5; high-tech elites 95–9; new identities 91–9; Thatcherite challenge to established 28–31
migration 142–3
Milton Keynes 66–7, 123, 125, 138; identity-formation 83–7; South Buckinghamshire and resistance to growth 87–8
Mohan, J. 25, 35, 43, 52
Moore, B. 127
Morgan, K. 22, 130
Morley, D. 86
mortgage debt 20
mortgage tax relief 21, 108
motherhood 95
Mott Report 78
multinational corporations 80, 81
Murdoch, J. 67, 84
Murray, R. 13, 120

narratives of growth 13–17
National Health Service 43, 101
negative equity 109, 120, 124
neo-corporatism 134
neo-liberalism 2; compared with Wilson modernization 53–4; EU regional policy 128; government actions 23–4; inequality 72–3, 139, 142–3; neo-liberal project 9, 12–13, 25; risk 112–13; spaces of 90, 91–9; spaces, places and times 138–43
nested hierarchies 60
networks: ethnic 106–7; of intermediate institutions 133–5; of privilege 16, 24
New Towns 66–7
New York 75, 76
Nordic countries 27–8
North American Free Trade Area (NAFTA) 129
north–south divide 2–3, 11, 32, 138–9

obsessionalism 98–9
Offe, C. 134
'one nation' politics 28, 91
openness, of the region 133
othering 82–9
overheating 120, 127
owner occupation 108–11, 113

paternalism *see* 'gentlemanly' capitalism
Peck, J.A. 118, 126
Phillips, M. 58
place: conceptualization of space and place 1–2; spaces, places and times of neo-liberalism 138–43
place-marketing 85–6
placed culture 29
Platt, J. 86
polarization, social 25–8
political power 11, 12
'post-Fordist' men 100–5
poverty 79; causes 26
power 11, 12, 33–4, 136; associationalism 133–4; uneven development and 51–2
private rented housing 111
private sector 85
privatization 23, 35, 52
privilege: networks of 16, 24
property: commercial 21–2, 120; residential 21; *see also* house prices, housing
Pryke, M.: changing nature of jobs in the security industry 103; globalization of markets 18, 22, 74; spatial bounds defined by City institutions 77, 78
public sector investment 23–4, 124
public services 52
'pull of Europe' 11, 12, 13
Putnam, R.D. 126

Quintas, P. 22, 77, 78

Rajan, A. 19
real crude gain 109
regeneration strategies 70–1
region: defining 1–2, 32–54
regional assemblies 134
regional authorities 130–1
regional class alliances 57
regional corporatism 132
regional expenditure: per capita 46
regional offices 132
regional policy 125, 126–9, 135–6
regional politics 129–32
regionalism 55
regulation, social 125–6, 129–32
relational approach 1–2, 4, 5; *see also* social relations
rented housing 108, 111
representation, system of 10
residential property 21; *see also* house prices, housing
resistance to growth 83–4, 87–9

RETI 132
Rhodes, J. 127
risk: home ownership 108–11; risk society 112–13
roads: expenditure on 45
Robins, K. 86
Rogers, J. 134, 135
Rubenstein, W.D. 16, 19–20
'Russian doll' approach 60
Rustin, M. 112

Sarre, P. 87, 110
Saunders, P. 109, 110
Savage, M. 21, 67, 74
Sayer, A. 22
scales, geographical 60
Schumpeterian workfare state 117
Scotland 11
Seavers, J.: boom and slump in property markets 20, 21, 108, 109; juxtaposition of place identities 87; motives of owner occupiers 110
security industry 103–5
SEEDS 122
SERPLAN 125
service industries 79–82; ethnicity 105–7; identity 90, 100–7; masculinity 100–5
'servile' class 27
sexual division of labour 100–1
Sibley, D. 83
skilled labour shortages 119
Slough 88
small firms 97
Smith, D. 11, 14, 20
social conditions 72–3
social inheritance 17
social polarization 25–8
social processes 34, 60
social regulation 125–6, 129–32
social relations 9, 17, 89, 137–8; components of growth and 24–5; between pro-growth and anti-growth areas 66–7, 87–9; London 67–8; and uneven development 50–2, 138–9; weakened between South East and rest of the country 138–9
South Buckinghamshire 55, 83–4, 87–9
space: conceptualization of space and place 1–2; globalization, masculinity, class and space 73–82; as product of social relations 50; spaces, places and times of neo-liberalism 138–43
St Mary's Island 71
Standard Region 33
Stones, R. 19
structural inequality 26
structured coherence 57–8, 60
Stubbings, F.H. 81
system of representation 10

tax reforms 21, 23, 35, 108
Thames Gateway initiative 71
Thatcherism 2, 3, 127, 141; and 'Englishness' 29–31, 105, 107; impact on Cambridge and City of London 77–8; narratives of growth 14–15
Thompson, G. 142
Thompson P. 101
Thrift, N. 21, 58, 68, 123; associationalism 132, 133, 134, 135; EU policy 128–9; institutional thickness 125–6; job growth 15, 19; 'sexy greedy' occupations 74, 93; Englishness 31, 105
Tickell, A. 118, 126
time 53–4; spaces, places and times of neo-liberalism 137–43
time–space distanciation 137–8
time–spaces: City of London 74–5; high-tech Cambridge 81–2
Tokyo 75, 76
tourism 33
Townsend, P. 25
trade: expenditure on 48
trade unionism 141
Training Enterprise Councils 131
transport: expenditure on 45; infrastructural investment 124
trickle-down effect 73
Turner, B.S. 112–13
'two nation' politics 25, 28
Tylor, P. 127

uncertainties 112–13
unemployment 123
unequal growth 118–22
uneven development 43–52, 73, 89, 135, 142–3; fragile growth 118–22; social relations and 50–2, 138–9

wages *see* incomes
Waldinger, R. 106
Webster, A.J. 76
Weiner, M.J. 12, 16
Weinstock, A. 141

white Englishness 29–31, 105–7, 140
'white ROSE' 29–31
Wield, D. 22, 77, 78
Wilkinson, F. 100
Wilson, H. 15, 29, 53, 127, 141
'winners and losers' principle 15
women: cleaning services 102; earnings distribution 36; in finance 94–5; and high technology 99; *see also* gender
work: globalization, class and 79–82; group identities 92–107; uncertainty 112–13; *see also* employment, labour market
work-poor 26–7
work-rich 26–7
World Bank 2

yuppie 74–5

Zukin, S. 67, 86